T0295086

# 3D/4D City Modelling

# 3D/4D City Modelling

## From Sensors to Applications

Edited by
Sultan Kocaman, Devrim Akca, Daniela Poli
and
Fabio Remondino

WHITTLES PUBLISHING

Published by
**Whittles Publishing,**
Dunbeath,
Caithness, KW6 6EG,
Scotland, UK
**www.whittlespublishing.com**



ISBN 978-184995-475-4



---

**Publisher's note**

Although our usual practice is to standardise UK/US spellings, in this book we have used both versions of some words, reflecting the international nature of its authorship. Readers of scientific literature will be familiar with the ise/ize variant and others but we believe this will not detract from the value of the content.

---



Printed and bound by CPI Group (UK) Ltd, Croydon, CR0 4YY

# Contents

# The contributors

## The Editors

Devrim Akca received both his BSc and MSc degrees at the Department of Geodesy and Photogrammetry Engineering, Karadeniz Technical University, in Trabzon, Turkey in 1997 and 2000, respectively, and his Ph.D. degree in Photogrammetry, at Swiss Federal Institute of Technology (ETH) Zurich, Switzerland in 2007. He worked as a senior research associate at the Institute of Geodesy and Photogrammetry (Geomatics), ETH Zurich between 2007 and 2009. Currently, he is a full Professor at the Department of Civil Engineering of Isik University, in Istanbul, Turkey. He received the ISPRS Best Paper by Young Authors Award, the Silver Medal of ETH Zurich, and the Carl Pulfrich Award.

Sultan Kocaman obtained her Ph.D. degree from ETH Zurich in photogrammetry and remote sensing, and has practical experience in the fields of cadastral surveying, GIS and spatial DBMS. She is currently a professor at Hacettepe University, Department of Geomatics Engineering, Ankara, Turkey. Besides photogrammetry and remote sensing, her research interests include the application of geospatial technologies and machine learning algorithms to various problem domains related to smart cities, natural hazards and land information management.

Daniela Poli received her first degree in Environmental Engineering at the TU Milan, Italy, in 1998, and her Ph.D. degree in Photogrammetry at Swiss Federal Institute of Technology (ETH) Zurich, Switzerland, in 2005. She worked as a senior research associate at the Institute of Geodesy and Photogrammetry (Geomatics), ETH Zurich (2007–2009) and at the Joint Research Center of the European Commission (2009–2012). Since 2012 she has worked for the aerial photogrammetric company AVT (Austria) and since 2021 manages the subsidiary in Italy.

Fabio Remondino graduated in Environmental Engineering from Politecnico of Milano in 1999 and received a PhD in Photogrammetry from ETH Zurich in 2006. He is now leading the 3D Optical Metrology research unit at FBK – Bruno Kessler Foundation, a public research centre located in Trento, Italy. His main research interests are in the field of reality-based surveying and 3D modelling, sensor and data fusion and 3D data classification.

## The Authors


Armin Gruen — Chair of Information Architecture, ETH Zurich Faculty of Architecture, Switzerland

Karsten Jacobsen — Institute of Photogrammetry and Geoinformation, Leibniz University Hannover, Germany

| | |
|---|---|
| Rongjun Qin | Associate Professor, Department of Civil, Environmental and Geodetic Engineering (CEGE), Department of Electrical and Computer Engineering (ECE), Translational Data Analytics Institute (TDAI) The Ohio State University, USA |
| Xu Huang | Postdoctoral researcher, Department of Civil, Environmental and Geodetic Engineering (CEGE), The Ohio State University, USA |
| Umit Isikdag | Professor of Construction Informatics, Department of Informatics, Mimar Sinan Fine Arts University, Turkey |
| Sisi Zlanatova | SHARP Professor, GRID, School of Built Environment, UNSW Sydney, Australia |
| Mehmet Buyukdemircioglu | Postdoctoral researcher, Department of Earth Observation Science, University of Twente, The Netherlands |
| Ken Arroyo Ohori | Postdoctoral researcher at the 3D Geoinformation Group, Faculty of Architecture and the Built Environment Delft University of Technology, The Netherlands |
| Hugo Ledoux | Associate Professor at the 3D Geoinformation Group, Faculty of Architecture and the Built Environment Delft University of Technology, The Netherlands |
| Jantien Stoter | Professor and Head of the 3D Geoinformation Group, Faculty of Architecture and the Built Environment Delft University of Technology, The Netherlands |
| Anna Labetski | Postgraduate researcher at the 3D Geoinformation Group, Faculty of Architecture and the Built Environment Delft University of Technology, The Netherlands |
| Stelios Vitalis | Postgraduate researcher at the 3D Geoinformation Group, Faculty of Architecture and the Built Environment Delft University of Technology, The Netherlands |
| Kavisha Kumar | Software Engieer, Bazel Build Architect, Veldhoven, The Netherlands |
| David Holland | Senior Research Scientist, Ordnance Survey, UK |
| Patrick Aeby | Federal Office of Topography, Wabern, Switzerland |
| Tobias Kellenberger | Federal Office of Topography, Wabern, Switzerland |
| Liam O'Sullivan | Federal Office of Topography, Wabern, Switzerland |
| Emanuel Schmassmann | Federal Office of Topography, Wabern, Switzerland |
| André Streilein | Federal Office of Topography, Wabern, Switzerland |
| Michael Zwick | Federal Office of Topography, Wabern, Switzerland |

# Chapter 1
# Introduction to 3D/4D city modelling

*Sultan Kocaman, Daniela Poli, Devrim Akca, Fabio Remondino and Armin Gruen*

The art of map-making has gone through revolutionary changes with the recent advances in geospatial technologies. The conventional 'sample (select) and measure' strategy has moved into 'first measure and then sample', such as the extraction of objects of interest from the millions of points in a point cloud. Professional surveyors receive significant support from non-professional users, as in the case of volunteer geographical information (VGI). Map presentation has evolved from 2D paper maps towards smartphone screens, web- and cloud-based platforms and preferably with 3D perception. Static geographical information is no longer the ultimate map product, when 2D and 3D spatiotemporal queries can be performed rapidly and efficiently with the help of spatial database management systems (DBMS) and thus new information and knowledge can be synthesized.

The concept of 3D and 4D (space and time) city modelling gains its grounds from these developments, where a city model actually becomes the main geometrical unit of a 3D geographical information system (GIS) environment. 3D spatial queries open up new horizons for many application areas, such as ecosystem modelling and simulations, integration of indoor data with the help of building information models (BIM), etc. Furthermore, semantic information associated with the 3D data ensures spatiotemporal querying and analysis. 3D city models and GIS empower the smart city concept and are inevitably crucial for the smart management of cities in which the majority of humans live.

The main scope of this book is to provide insights into the different generation stages and the system components of city models. Chapter 2, 'Sensor and data acquisition', focuses on data collection methods with high-resolution optical satellite imagery. Although various other platforms (such as aerial and terrestrial mobile) and sensors (such as multispectral cameras mounted on unmanned aerial vehicles (UAVs), smartphone cameras, synthetic aperture radar (SAR), etc.) exist, it is almost impossible to cover the whole range of data collection methods. Therefore, specific emphasis is given to data production with optical satellites, as they can provide continuous geometric data at the city scale with the increasing number of satellites launched by many nations and organizations; and semantic data extraction is relatively easy with advanced image classification techniques and the availability of a wide range of spectral bands.

Chapter 3, 'Geometric processing for image-based 3D object modelling', mainly covers accurate automated modelling approaches with high performance and low computational cost. State-of-the art texture mapping methods are also covered in the modelling process as they ensure more complete and realistic models with implicit semantic information, i.e. visual clues on the models.

Chapter 4, 'Utilizing BIM as a resource for representation and management of indoor information', provides a broad overview of BIM (Building Information Modelling) standards and their functionality. The opportunities offered by BIM and the application areas are quite broad and, despite the promising possibilities, the integration of BIM with GIS remains an emerging research field for the time being.

Chapter 5, 'A review of existing tools and methods for the management and visualization of 3D city models', presents the current status of existing tools and software for the management and presentation of city models, with a special focus on their capabilities and the main problems. Although a fully automated one-button procedure to generate city models is not yet available, several solutions to sub-components exist and the interoperability between these solutions is relatively high thanks to collaborative efforts of OGC (Open Geospatial Consortium) members. On the other hand, the storing or dynamic access and querying of semantic data and dynamic re-modelling (or geometry modification) after changes to object geometries, seem to be major issues and require more effort.

Chapter 6, 'Representations of 3D and 4D city models', covers the different algorithmic approaches and data formats for the models and the interoperability between them. Topology and generalization topics are also covered in this chapter, and different methods are described and discussed accordingly. The chapter concludes with a brief discussion on 4D visualization.

Chapters 7 and 8 focus on change detection approaches for the purpose of city model updates. Although a large number of change detection methods is available in 2D, the methods for 3D are relatively new although their number is increasing. Applicable automated change detection approaches are essential for large mapping organizations and national mapping agencies, since up-to-date geographical and thematic data as well as big geospatial data remain mainly their responsibility.

The final chapter of the book, 'Topographic landscape model of Switzerland swissTLM$^{3D}$', demonstrates a state-of-the-art application of city models at country level. Specific details on the city model production process, such as geometric de-generalization, data densification and adding the third dimension, are provided with a brief lessons-learned section and future perspectives. Different aspects of quality requirements, such as authority, accuracy, consistency, correctness etc., have been provided in the vision of 3D cadasters.

Overall, the main issues and open research areas based on the book contents may be listed as:

- A system (GIS) approach to city models is essential for the efficient management and utilization of the models if the vision of creating and simulating a digital earth is to be realized, thus integrating human and ecosystem aspects integrated to the models, providing solutions to environment-related and smart city problems, and supporting the 2030 Sustainable Development Goals of the United Nations to ensure a higher quality of living.

- Although city models can be generated with semi-automated procedures, the level of automation can be increased with the utilization of data from newer sensors, improved integration of existing sensor data, and the use of novel machine learning methods.
- Semantic information extraction and management approaches are active research areas that need further development. Standardized data structures and spatial DBMS solutions are required for the efficient storage and updating of geometry together with semantic data, which allow high-performance web-based visualization at the same time.
- Although there are several standards for data storage and exchange, such as CityGML, no standards or consensus on the objects contents of city models is yet available. City models are not only composed of buildings but may contain every geo-related object (e.g. trees, underground infrastructure, flora, etc.). With the increasing variety of model details, more studies on better standards and spatial data infrastructures would reduce data interoperability problems and the complexity of data mining in the future.
- The integration of user-generated geographical data and the handling and analysis of the big geodata, are probably only possible with the wider use of machine learning algorithms and artificial intelligence approaches, i.e. Geo-AI.

Finally, it may be said that 3D/4D city modelling studies will form the future of map-making.

# Chapter 2
# Sensors and data acquisition

*Karsten Jacobsen*

## 2.1 Very high resolution optical satellite sensors

### 2.1.1 Sensors

Sputnik – the first satellite – was launched by the Soviet Union in October 1957, thus starting the race into space. The USA followed with the satellite Explorer in January 1958. During the Cold War, the military were eager to have images of foreign countries. The USA used the high-altitude U2 aircraft to fly over the Soviet Union; the aircraft was shot down in May 1960, ending this phase of military reconnaissance. Nevertheless, in 1959 the USA began development of the KH-1 program, later renamed to CORONA, using a satellite with a panoramic camera returning a capsule with exposed film. The first successful launch was in August 1960, marking the first real application of a satellite. The ground resolution of the first CORONA panoramic images was in the range of 7.5 m, improved to 1.8 m in 1960 (Doyle, 1996). With the KH-4 program a second camera gave the possibility of stereoscopic coverage (Schneider *et al.*, 2001). The images from the CORONA program taken up to 1963 have been declassified and are available for a handling charge. The Soviet Union followed the USA with a series of space cameras, such as the KFA-1000, MK4, KFA-3000, KVR-1000 and TK350, which have been returned from space intact for re-launch.

The first space images available for civilian application were Landsat images in 1972. However, with a 79 m ground sampling distance (GSD) they are useful for classification purposes but not for mapping. Under GSD we understand the distance of neighboring pixel centers projected on the earth. GSD is a precise definition and does not depend on resampling of the projected pixel size. The application of space images for civilian mapping started with the launch of French SPOT-1 in 1986. Nevertheless a 10m GSD for the panchromatic band is not optimal for mapping purposes. The real breakthrough of satellite imaging came in 1999 with IKONOS, having 0.81 m GSD in the panchromatic band for nadir images. For civilian or multiple use, up to May 2017 a total of 39 optical satellites, with 1m GSD and smaller for the panchromatic band, have been launched successfully by governmental and private organizations from 14 countries. Today there is an open market for the construction and launch of such satellites.

The huge amount of image data is reduced by a lower resolution for the colour bands. As standard linear geometric relation between the panchromatic and the colour bands the factor 4.0 is used. This corresponds to the sensitivity of the human eye, which has approximately 4.5 million cones, sensitive for colour, and 90 million rods, sensitive for grey values, corresponding to a relation 1:20 or linear 1:4.5. By pan-sharpening, the lower resolution colour bands are merged with the higher resolution panchromatic bands to provide high-resolution colour images. The original meaning of panchromatic corresponds to the range visible for the human eye but, for satellite images, it is used for the grey values covering a different spectral range, which often is extended to the near infrared. Extending the range to the near infrared has advantages for the image matching of forest areas, which are dark in the visible range but bright in the near infrared. The pan-sharpening is more complicated if the spectral range is not identical to the colour bands for blue, green and red; nevertheless there are some solutions for this.

Satellites are usually classified as in Table 2.1. With technical development, satellite components are getting smaller and satellites are getting lighter, so even cube-satellites (10 cm × 10 cm × 10 cm) and triple cube-satellites, of just a few kilograms, are in use for optical imaging.

Optical satellite cameras are classified by their ground resolution and the number and range of spectral channels. The classification is not unequivocal; it depends partly upon the group of users, but the classification shown in Table 2.2 is widely used and will be used here. As mentioned before, colour bands usually have a lower resolution; this lower resolution is not respected for the grouping of optical satellites.

Most satellite cameras used for mapping purposes have a panchromatic channel and some colour channels as for red, green, blue (RGB) and near infrared (NIR). Only WorldView-2 and -3 have eight spectral bands with four times the GSD as the panchromatic band. Some cameras also have an imaging sensor in the short wave infrared range (SWIR), but with a lower resolution due to reduced reflected energy in this spectral range. Cameras with separate CCD- or CMOS-lines for the used spectral ranges are called multispectral sensors, while cameras with several colour bands, usually equipped with a prism to separate the received light energy into a high number of spectral bands, are called hyperspectral cameras. Hyperspectral sensors may have from 20 to 640 channels, but the geometric resolution is limited due to the limited energy in the very narrow bands and the very high data volume. They are mainly used for classification of vegetation and/or minerals.

The spectral range for the different colour bands and the panchromatic band shows some variations for optical satellites (Table 2.3).

**Table 2.1** Classification of satellites according to their weight.

| Small satellites | | | | Large satellites |
|---|---|---|---|---|
| Pico | Nano | Micro | Mini | |
| ≤1 kg | >1–10 kg | >10–100 kg | >100–1000 kg | >1000 kg |

**Table 2.2** Classification of optical satellite cameras according to panchromatic GSD.

| Low resolution | Medium resolution | High resolution | Very high resolution |
|---|---|---|---|
| >30 m | >5 m up to ≤30 m | >1 m up to ≤5 m | ≤1 m |

**Table 2.3** Variation and range of spectral bands of very high resolution satellites (μm).

| Band | Panchromatic | Blue | Green | Red | NIR |
|---|---|---|---|---|---|
| from | 0.45/0.51 | 0.43/0.44 | 0.50/0.52 | 0.59/0.63 | 0.74/0.85 |
| to | 0.65/0.90 | 0.51/0.55 | 0.58/0.62 | 0.68/0.71 | 0.85/0.94 |

All optical satellites are equipped with sensors for the determination of attitude and position. Position can be determined in the sub-decimeter range, while the attitude is limited by the direct sensor orientation – i.e. the location accuracy of the images projected to the ground. The attitude determination is based on a combination of stellar cameras and gyros. The gyros have only a good relative accuracy – the drift has to be determined by the stellar cameras. This is complicated by the permanent rotation of the satellites. The first satellites had cameras that could be rotated against the satellite body or had at least a rotating mirror in front. Today, body pointing satellites are dominant – the main part of the satellite is the camera, which is pointed by rotating the whole satellite. The rotation may be based on reaction wheels – gyros strapped down to the satellite – or control moment gyros, fixed in the inertial space. Control moment gyros allow a faster satellite rotation. If the gyros are accelerated, a moment goes to the satellite and the satellite rotates based solely on the electric energy from the solar panels. The speed of the view direction change – the slewing speed – is an important factor for the economy of the satellite. The tasking of the satellite to take images of the ordered areas, supported by cloud cover information, allows the taking of more images if the view direction can be changed quickly from one area to the next. The required time for changing the viewed area by a ground distance over 200 km and being ready for taking the next image – the so called slewing (Table 2.4) – depends upon the angular speed of changing the view direction and the satellite elevation. Nevertheless, the continuous acceleration and the subsequent slowing down of the view direction change brings with it the attendant problem of satellite vibration, which has to be avoided if a better imaging quality is to be achieved and geometric stability guaranteed.

The slewing time for QuickBird was so slow that no additional images in addition to a stereo combination could be taken. Also IKONOS and GeoEye have a limited slewing time due to the use of reaction wheels. This influenced the economic situation of the former company GeoEye so that, finally, it was acquired by DigitalGlobe.

Only the new small satellites are equipped with CCD or CMOS frames, the other optical satellites having a combination of CCD-lines or transfer delay and integration

**Table 2.4** Required time for slewing over 200 km ground distance.

| | IKONOS | QuickBird | WV-1 | GeoEye | WV-2 | Pleiades | WV-3 | WV-4 | TeLEOS-1 |
|---|---|---|---|---|---|---|---|---|---|
| Δt [sec] | 18 | 37 | 10 | 20 | 10 | 11 | 12 | 11 | 14 |
| gyros | RW | RW | CMG | RW | CMG | CMG | CMG | CMG | CMG |

WV = WorldView; RW = reaction wheels; CMG = control moment gyros arrays (TDI). The transfer of the charge within a long CCD-line to the read out register is too time consuming for very high resolution satellites. The satellite speed in the orbit is approximately 7.6 km/sec in the flying height range used for optical satellites. For 0.31 m GSD only 0.04 msec are available for taking an image line if the view direction of imaging is not changed in the flight direction. For this reason a combination of shorter CCD-lines is required. The CCD-lines cannot be put together without gaps: they have to be arranged in staggered mode (Figure 2.1). The colour CCD-lines are separated from the panchromatic CCD-lines; few optical satellites have two arrangements of colour CCD-lines – one for the forward scanning mode and one for the backward scanning mode.

By theory a correct merge of all sub-images is only possible in an object plane due to the different projection centers t1 up to t4 corresponding to the imaging instants (Figure 2.2). In reality the angular difference of sub-images is not as large as shown in Figure 2.2. For most optical satellites the merging causes errors of one panchromatic pixel for height differences against the reference plane of 500 m up to 4 km, if no beam splitters are used, as in case of Pleiades, avoiding such problems.

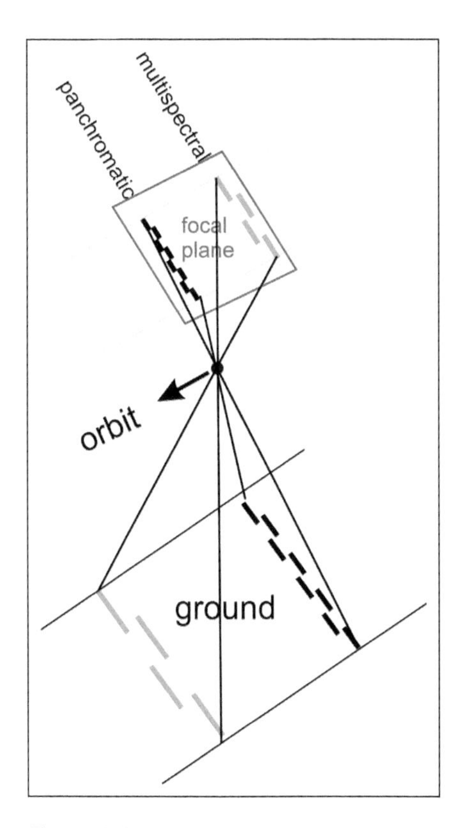


Figure 2.1 Combination of CCD-lines in the focal plane and corresponding ground location.

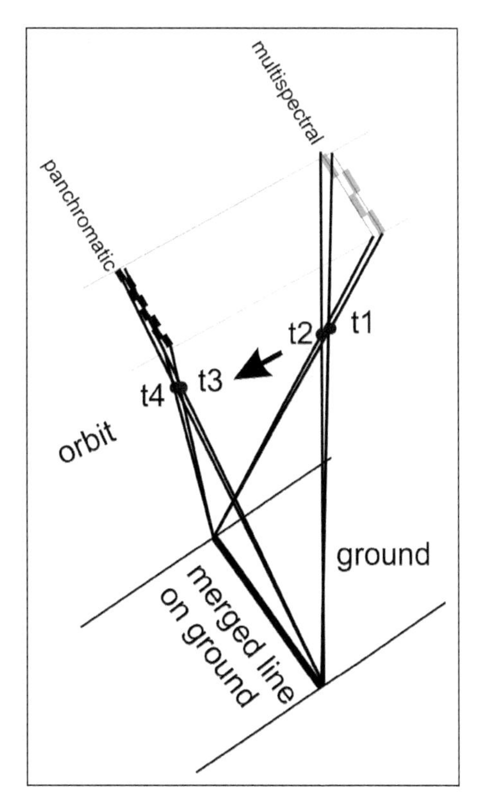


Figure 2.2 Combination of sub-images merged in object space and corresponding projection centers t1 – t4.

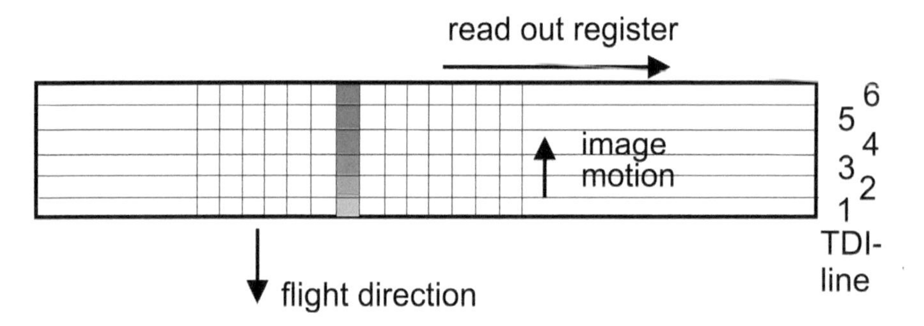


Figure 2.3 Charge transfer in a TDI sensor corresponding to image motion.

The very short exposure time – 0.04 msec for images with 0.31 m GSD – does not lead to images with a satisfying signal-to-noise ratio. This problem can be solved by TDI sensors. The charge generated in TDI-line 1, corresponding to the image motion speed, is shifted to TDI-line 2 where more charge is generated and then shifted to TDI-line 3 and so on, up to the specified number of TDI-lines (Figure 2.3). By this means a satisfying charge can be accumulated leading to a low signal-to-noise ratio. The number of lines in a TDI sensor is up to approximately 60 lines. The number of effectively used TDI-lines depends upon the object brightness but is limited by the cross-orbit nadir angle.

Nearly all Earth observation optical satellites have a sun-synchronous orbit imaging approximately at the same local time for each satellite with an incidence angle of approximately 96.5° for 250-km flying height and 98.2° for 700-km flying height. The imaging is usually in the descending orbit; only South Korea prefers an ascending orbit, due to the shape of the country. The time for equator crossing is generally between 10:00 and 11:00 o clock; only India prefers an earlier time. The time for equator crossing should not be too early, for reaching a satisfying sun elevation, and it should not be too late, to avoid clouds building up during the day. However, for the current groups of satellites later equator crossings are also used.

The cameras mounted on the first optical satellites had a fixed orientation during imaging. SPOT-5 changed the mirror orientation continuously to reach a north-south scene limit. Today, most high- and very high resolution optical satellites have the ability to take the scenes in more directions (Figures 2.4 and 2.5), even against the flight direction; that means, they need a permanent rotation during imaging. This permanent rotation is used also for reaching the required imaging time in an asynchronous mode. The path length of the satellite during imaging in the orbit can correspond to the footprint distance as in the synchronous case (S) in Figure 2.4, but the scene can also be scanned during a longer time, corresponding to asynchronous imaging (AS). It is also possible to have asynchronous imaging with a shorter length in the orbit, as is the case for WorldView-1 (Table 2.5), although it can be extreme as is the case for EROS-B with a slow-down factor of 3.3 – i.e. it takes 3.3 times as much time for imaging as for passing the scene area. The large slow down factor is not a problem for the individual scene, but it reduces the imaging capacity.

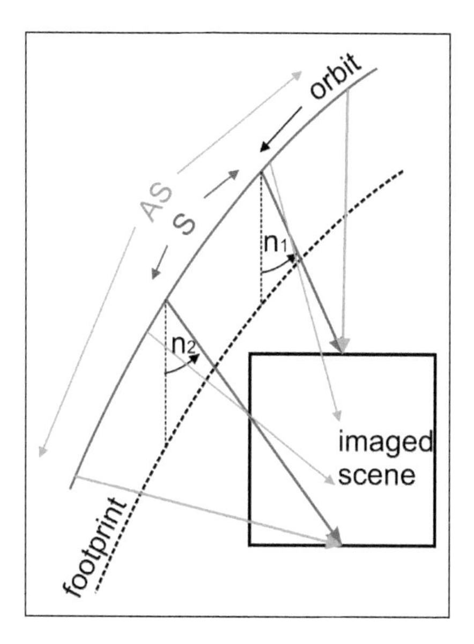


Figure 2.4 Synchronous (S) and asynchronous (AS) imaging.

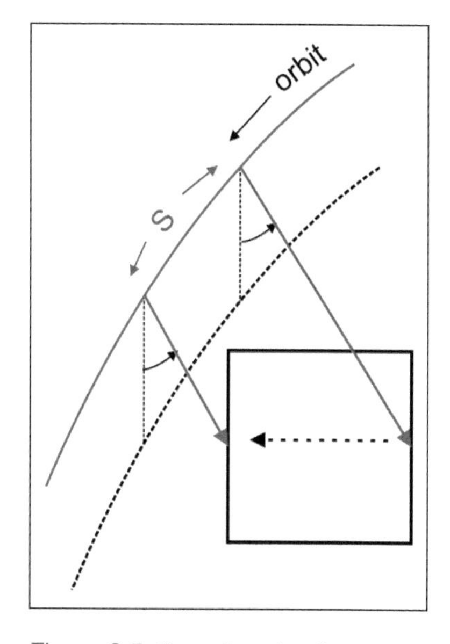


Figure 2.5 Scene imaging from east to west.

**Table 2.5** Slow down factor of asynchronous imaging for a selection of optical satellites.

| Satellite | Sampling rate (lines/sec) | Footprint speed (km/sec) | Nominal GSD (m) | Slow down factor |
|---|---|---|---|---|
| IKONOS | 6500 | 6.79 | 0.82<br>1.00 | 1.27<br>1.04 |
| QuickBird | 6900 | 6.89 | 0.61 | 1.64 |
| WV-1 | 24000 (only pan) | 7.06 | 0.50 | 0.59 |
| WV-2 | 24000 pan only<br>12000 pan + ms | 6.66 | 0.50 | **0.55**<br>1.11 |
| GeoEye-1 | 20000 pan only<br>10000 pan + ms | 6.79 | 0.50 | 0.68<br>1.36 |
| Cartosat 2 | 2732 | 6.85 | 0.80 | 3.13 |
| Kompsat-2 | 7100 | 6.78 | 1.0 | 0.96 |
| Pleiades | 14000 | 6.77 | 0.5 | 0.97 |
| EROS-B | ≤3050 | 7.04 | 0.7 | **3.30** |

ms = multispectral

Also in case of synchronous imaging of a scene with parallel limits to the ground coordinate system (usually UTM) the view direction at the start of the scene ($n_1$) is not the same as at the end of the scene ($n_2$) due to the inclined orbit not being exactly in north-south direction.

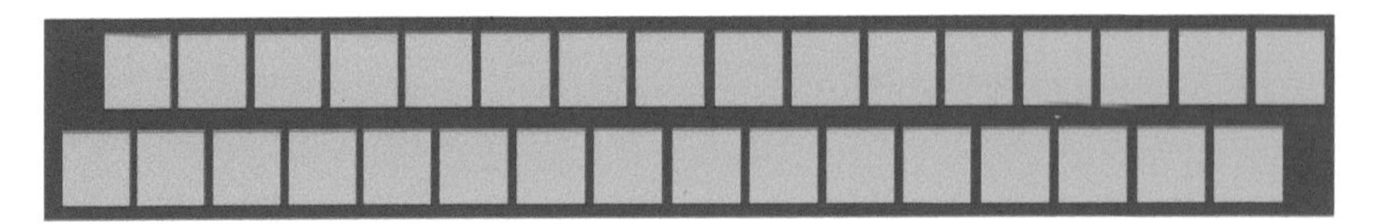

Figure 2.6 Staggered CCD-lines.

For a satisfying signal-to-noise ratio a TDI sensor or a strong slow down factor is required for sensors with a GSD of 1 m and less. Another possibility is the use of staggered CCD-lines (Figure 2.6). Staggered CCD-lines have in line direction two lines shifted against each other by half a pixel in line direction. The sampling rate corresponds to half a pixel, so an image with a grid of pixels overlapping 50% in line and in sample direction is generated – or in other words 50% over-sampling in line and sample direction exists. The GSD of a staggered CCD-line corresponds to half a pixel size projected to object space. Of course, with this technique the same image quality as based on usual CCD-lines is not possible. An investigation of the quality of satellite images (Jacobsen, 2011) based on edge analysis showed that images with staggered sensors have a factor for effective resolution of approximately 1.2. That means the nominal GSD has to be multiplied by this factor to get an effective GSD corresponding to a GSD based on sensors with such an original ground resolution. Or in other words, staggered images with 1 m GSD have the quality of a sensor with a single CCD-line of 1.2 m. This is still quite a lot better as the projected pixel size of 2 m. For example, OrbView-3, Cartosat and SPOT-5 supermode are/were based on staggered CCDs.

The number of optical satellites and very high resolution satellites is growing. Table 2.6 lists over 50 satellites, including the three DMC-3 satellites, Skysat-3 up to -7 and the two GaoJing-1. The images of these optical satellites are available for civilian use. In addition there exists a higher number of military satellites but these are not discussed here. During the first 8 years, 9 satellites have been launched, and during the last 12 years 48 in total. The number of countries operating very high resolution satellites is growing, which is partly due to the fact that the components of satellites (or even whole satellites) can be bought. With satellite launching, there is international competition. There is a trend to higher ground resolution, which has now reached 0.31 m. SkySat-1 and -2 are included in Table 2.6 even if the ground resolution is 1.1 m; however, with a weight of just 120 kg they reflect the trend towards smaller satellites, which is continued with Skysat-3 to -7. The company Planet has a configuration of 3-Cube-Sats (10 cm × 10 cm × 35 cm) under the name Flock, later named Dove, weighing just 5.8 kg. Currently 120 Dove satellites are active. EarthNow, supported by Airbus DS, Bill Gates and the Japan Soft Bank Group, are planning a constellation of 500 earth observation satellites capable of delivering on-line videos from space. Furthermore, China is establishing a fleet of small earth observation satellites. The SkySat, now owned by Planet, has a combination of 3 slightly overlapping CMOS-arrays of 2560 × 2160 pixels. 50% of the arrays are used for the panchromatic channel and 12.5% have a blue filter, 12.5%

**Table 2.6** Very high resolution optical satellites.

| Satellite | Launch | GSD (m) | Swath (km) | Country | Satellite | Launch | GSD (m) | Swath (km) | Country |
|---|---|---|---|---|---|---|---|---|---|
| *IKONOS* | *1999* | *0.81* | *11.3* | *USA* | KazEOSat-1 | 2014 | 1 | 20 | Kazakhstan |
| *QuickBird* | *2001* | *0.62* | *16.5* | *USA* | SkySat-2 | 2014 | 1.1 | 8 | USA |
| *TES* | *2001* | *1* | *13* | *India* | WorldView-3 | 2014 | 0.31 | 13.1 | USA |
| *OrbView-3* | *2003* | *1* | *8* | *USA* | GaoFen-2 | 2014 | 0.81 | 45 | China |
| EROS-B | 2006 | 0.82 | 7 | Israel | Resurs-P2 | 2014 | 1 | 36 | Russia |
| Resurs-DK1 | 2006 | 1 | 28 | Russia | ASNARO-1 | 2014 | 0.5 | 10 | Japan |
| Kompsat-2 | 2006 | 1 | 15 | S. Korea | DMC-3 / Triple | 2015 | 1 | 23.4 | UK/China |
| Cartosat-2 | 2007 | 1 | 10 | India | Kompsat-3A | 2015 | 0.55 | 13.2 | S. Korea |
| WorldView-1 | 2007 | 0.46 | 17.6 | USA | Jilin-1 Opt.-A | 2015 | 0.72 | 11.8 | China |
| GeoEye-1 | 2008 | 0.46 | 17.1 | USA | TelEOS-1 | 2015 | 1 | 12 | Singapore |
| Cartosat-2A | 2008 | 1 | 10 | India | Resurs-P3 | 2016 | 1 | 36 | Russia |
| WorldView-2 | 2009 | 0.41 | 14.5 | USA | Nu-Sat-1, -2 | 2016 | 1 | 5×5 | Argentine |
| Cartosat-2B | 2010 | 1 | 10 | India | PeruSat-1 | 2016 | 0.70 | 20 | Peru |
| Pleiades-1A | 2011 | 0.5 | 20 | France | Cartosat-2C | 2016 | 0.65 | 8 | India |
| GaoFen | 2012 | 0.8 | 28 | China | SkySat-3 -7 | 2016 | 0.9 | 7 | USA |
| Yaogan 14 | 2012 | 0.8 | 23 | China | *WorldView-4* | *2016* | *0.31* | *13.1* | *USA* |
| Kompsat-3 | 2012 | 0.7 | 16.8 | S. Korea | GaoJing-1, -2 | 2016 | 0.5 | 12 | China |
| Pleiades-1B | 2012 | 0.5 | 20 | France | Göktürk-1 | 2016 | 0.7 | 20 | Turkey |
| Resurs-P1 | 2013 | 1 | 36 | Russia | Cartosat-2E | 2017 | 0.65 | 10 | India |
| DubaiSat-2 | 2013 | 1 | 12 | UAE | Mohammed VI | 2017 | 0.70 | 20 | Morocco |
| SkySat-1 | 2013 | 1.1 | 8 | USA | Cartosat-2F | 2018 | 0.65 | 10 | India |
| Deimos-2 | 2014 | 0.75 | 12 | Spain | KhalifaSat | 2018 | 0.75 | 12 | UAE |
| UrtheCast IRIS | 2014 | 1 | 4 | Canada | FalconEye-1 | 2019 | 0.70 | 20 | UAE |

italic = off-duty

a green filter and the same for red and NIR to enable colour images. Based on the arrays, videos can also be generated. Skysat, Dove and the other small satellites are made for visualization not so much for mapping purposes (d'Angelo, Kuschk and Reinartz, 2014).

The GSD listed in Table 2.6 correspond to the delivered images; they must not be the same as the projected pixel size as mentioned above due to staggered CCD-lines. Another effect may be enlargement during the image processing. For example, Pleiades-1A and -1B have a physical GSD of 70 cm but they are delivered with 50 cm GSD. Jacobsen *et al.* (2016) analyzed the radiometric image quality of Pleiades-1A images in relation to other high- and very high resolution space images. The factor for effective resolution, based on edge analysis, for the Pleiades with 0.5 m GSD was between 0.9 and 1.0, having a better quality than most other analyzed space image types. Also the signal-to-noise ratio and the blur coefficient is satisfying. A visual comparison between Pleiades images with 50 cm GSD and WorldView-1 images with 56 cm GSD and QuickBird with 62 cm GSD gave clearly better results for Pleiades, meaning that the zooming of Pleiades images did not cause any problems.

Not all images from the satellites in Table 2.6 are commercially available. Some of these satellites are predominantly for the countries owning the satellites. Russia is not distributing images with better than 2 m GSD. Nevertheless a significant number of images can be ordered and satisfying competition exists. The major international satellite image distributors are Digital Globe, Airbus DS, the Indian ISRO, the South Korean KARI and the Japanese JAXA.

With Pleiades Neo we have more satellites with approximately 30 cm GSD. The other satellite image distributors are also planning additional very high resolution earth observation satellites. Due to miniaturization of the components and reductions in price, the trend is towards satellite fleets.

### 2.1.2 Image geometry and calibration

Any sensor line has a different exterior orientation due to permanently changed projection center and rotation. Nevertheless the changes are small and should be continuous within the scene size, allowing a handling of the images as one unit if no sensor vibration occurs during imaging. In addition the sensor lines are based on a combination of different CCD-lines that should fit together and should be aligned and scaled after calibration. The TDI has no direct influence on the geometry; it is only if the number of used lines for charge integration is changed that the relation between panchromatic and multispectral lines is affected. The inner orientation of the satellite camera is calibrated in the laboratory before launch and in orbit at regular time intervals, due to possible change caused by launch shock, drying out in space and thermal influences. The standard user has no access to the original sub-images and only gets ready merged images based on actual calibrations.

The author had the possibility of an in-orbit calibration of IRS-1C images (Jacobsen, 1997). IRS-1C has had 3 CCD-lines delivering 3 sub-images. Based on a configuration of three full images, each taken with 1-day time interval, with one nadir view and one view with 19° off-nadir angle from west and 20° off-nadir angle from east, the sub-images were well overlapping and could be joined together by tie points and oriented with few ground control points (GCPs). By block adjustment using the orbit information, supported by the view direction, a common adjustment of the relation of the sub-images together with the sensor orientation was possible with specially adapted additional parameters fitting the known sub-image geometry. The special additional parameters have to shift, rotate and scale the outer CCD-lines in relation to the center CCD-line, which is used as reference. Within the homogenous CCD-lines a change in the geometry against the pre-calibration is not expected and could not be seen. According to theory 5 GCPs satisfying the integrated adjustment together with the tie points, but for better reliability 9 GCPs have been used. Schröder, Müller and Reinartz (1999) describe the calibration of the MOMS-sensor, Valorge *et al.* (2004) the calibration of SPOT, and Kocaman and Grün (2007) the calibration of ALOS/PRISM.

Usually the user is getting only homogenous images based on the calibrated inner orientation; this limits the geometric check to a validation of the scene geometry. A single scene can be validated only against GCPs but a stereo scene allows validation by relative orientation and via a generated height model. The direct sensor orientation is determined by global navigation satellite systems (GNSS) and the combination of stellar sensors and gyro information. Today, GNSS may be based on the American GPS, the Russian GLONASS, the European Galileo and the Chinese Beidou. The satellite position can be determined with a standard deviation of 5 cm but the attitudes are the limitation. The accuracy of the direct sensor orientation has improved over the years (Figure 2.7). The figures from the satellite operating companies are realistic.

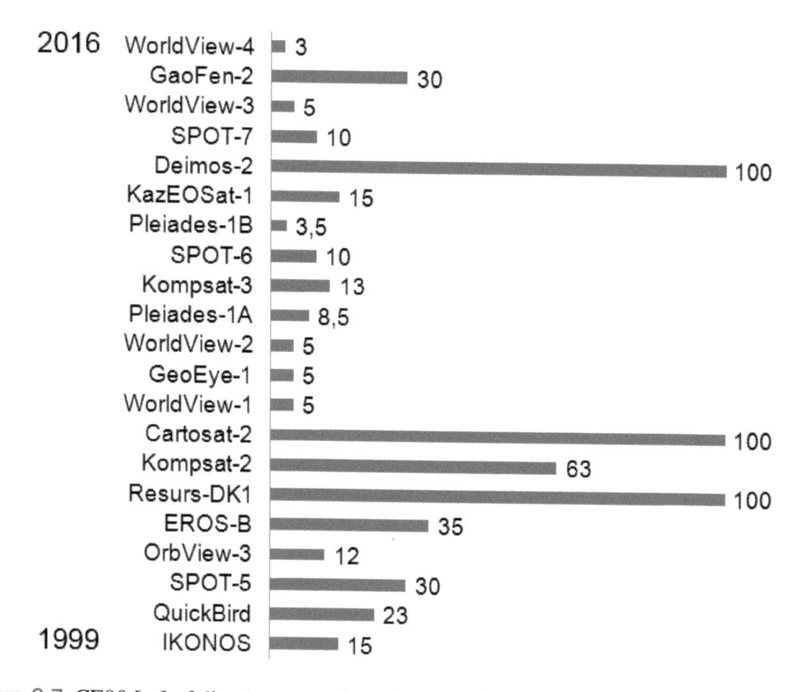


Figure 2.7 CE90 [m] of direct sensor orientation, sorted by launch time.

In the case of same accuracy for *X* and *Y* and normal distributed data, the circular error CE90 corresponds to the standard deviation for *X* (SX) or *Y* (SY) coordinates multiplied by the factor 2.146. Or reverse 3 m CE90 correspond to SX = 3 m/2.146 = 1.40 m. This accuracy, reached by WorldView-4, allows sensor orientation without GCPs that can be used for several purposes; often, however, more problems exist with the local Geodetic Datum as with the direct sensor orientation. The Pleiades type satellites, as PeruSat-1, Khalifa and Mohammed VI-A, constructed by Airbus DS, correspond to Pleiades 1B. To attain full accuracy of object point determination with satellite images, and for better reliability, orientation with GCPs is required.

The classical satellite image orientation based on orbit and attitude information requires a knowledge of these parameters, but today in most cases only the replacement model by rational polynomial coefficients is available and used (Grodecki, 2001; Grodecki and Dial, 2003).

$$x_{ij} = \frac{P_{i1}(X, Y, Z)_j}{P_{i2}(X, Y, Z)_j} \quad y_{ij} = \frac{P_{i3}(X, Y, Z)_j}{P_{i4}(X, Y, Z)_j} \tag{2.1.2.1}$$

rational polynomial coefficients, where $x_{ij}$, $y_{ij}$ = normalized image coordinates

$$\begin{aligned} P_{i1}(X, Y, Z)_j = {} & a_1 + a_2 \cdot Y + a_3 \cdot X + a_4 \cdot Z + a_5 \cdot Y \cdot X + a_6 \cdot Y \cdot Z \\ & + a_7 \cdot X \cdot Z + a_8 \cdot Y^2 + a_9 \cdot X^2 + a_{10} \cdot Z^2 \\ & + a_{11} \cdot X \cdot Y \cdot Z + a_{12} \cdot Y^3 + a_{13} \cdot Y \cdot X^2 \\ & + a_{14} \cdot Y \cdot Z^2 + a_{15} \cdot Y^2 X + a_{16} \cdot X^3 + a_{17} \cdot X \cdot Z^2 \\ & + a_{18} \cdot Y^2 \cdot Z + a_{19} \cdot X^2 \cdot Z + a_{20} \cdot Z^3 \end{aligned} \tag{2.1.2.2}$$

coefficients of third-order polynomials used for RPCs, where *X*, *Y*, *Z* = normalized geographic coordinates

The RPCs (equation 2.1.2.1 and 2.1.2.2) give the relation between the image coordinates $x_{ij}$, $y_{ij}$ and the geographic object coordinates (*X*, *Y*, *Z*). The coordinates are normalized to values between 0.0 and 1.0 by splitting the lowest coordinate of the scene range and dividing the remaining through the coordinate range. This avoids numeric computational problems. The coefficients are adjusted based on a cube of points in the object space covered by the image. The image coordinates are calculated for this artificial cube with given *X*, *Y*, *Z* coordinates based on the direct sensor orientation, using the geometric sensor model. Two coefficients have to be fixed to avoid indefinite results – usually $a_1$ in the polynomials $P_{i2}$ and $P_{i4}$ are set to 1.0.

The 80 RPC coefficients identify a rational model, that the model the real geometric situation. With the relation of third-order polynomials within one scene, the geometric reference is given without using the real geometric model, which is different from satellite camera to satellite camera. This standardized orientation model under usual conditions is delivering the same accuracy as the real geometric model (Büyüksalih and Jacobsen 2005). Only if neighboring scenes of the same flight line are merged together may the RPCs for such extended scenes not be accurate enough. Third-order RPCs are also unable to express the influence of higher order satellite jitter, even if this may be recorded by gyros. In this case only a change of image geometry can help.

The RPCs should also include the refraction influence. The refraction for satellites is caused by the atmosphere, which is denser close to the object, so that under usual conditions its influence is clearly below 1 m for larger nadir angles.

Standard RPCs describe the image coordinates as a function of the ground coordinates. Within the orientation program and also the program for determining the ground positions based on intersection, the reverse transformation is required too. Only in case of Pleiades images are the reverse RPCs provided. In general this is not a problem due to an iterative back-projection of the RPCs. The RPCs use geographic ground coordinates, but usually the ground control points are available in national net coordinates; mostly UTM is used. Thus, orientation programs should have the capability of a transformation from geographic coordinates to national net coordinates and reverse.

RPC orientation without use of ground control has the accuracy of the direct sensor orientation. For accurate handling of the satellite images this is not satisfactory, requiring support by GCPs. The bias-corrected RPC orientation uses a two-dimensional correction in image or object space. As standard a two-dimensional affine transformation is used (equation 2.1.2.3), but it is possible to use statistical checks for the individual parameters of the affine transformation to identify whether all parameters are required. This may lead to a reduction in the bias correction by shifts in the *X*- and *Y*-direction (parameters A1 and A4 in equation 2.1.2.3).

$$X' = A1 + A2 * X + A3 * Y \qquad (2.1.2.3)$$

2-D affine transformation

$$Y' = A4 + A5 * X + A6 * Y$$

The affine transformation depends upon the GCPs used. The main problem is not the accuracy of the ground coordinates of the GCPs: the main problem is the identification of the GCPs in the images. A targeting of the GCPs before flight is the rare exception: usually objects which can be identified are used; this is simpler in urban than in rural areas. Symmetric points should be preferred due to the shift of corner points depending upon the illumination. Even by changing the screen brightness, shifts in corner points appear. In extreme cases this may reach a GSD. Problems also may occur if the ground location of the GCP is not flat. Different view directions from the orbit to the ground may cause shifts in the identified three-dimensional GCPs.

Under operational conditions a limited number of GCPs is used. They should be located at least close to the scene corners to obtain a satisfactory accuracy of the affine parameters. In addition an over-determination is required to identify errors of the GCPs. Nevertheless identification problems may influence the affine parameters. This has a limited influence for single scenes, as used for the generation of orthoimages, but, if due to the three-dimensional shape of the GCPs or uncertain identifications different shifts in the scenes of a stereo pair occur, the affine parameters of the stereo pair may be influenced in a different manner. Scale differences in base direction are causing a tilt of generated height models.

The classic satellite image orientation used the satellite positions and attitude information, available for certain time intervals. The satellite path interpolation is simple if an inertial ellipse is fitted to the positions; an approximation with polynomials should be avoided. In addition some additional parameters, partially corresponding to the bias correction, have to be used for optimal fit to the GCPs. For IKOOS, in the initial years no detailed orientation information was available – the operating company liked to earn additional money with precise orientations. However, the satellite elevation and azimuth for the scene center and the scene center position was included in the header data. With a rotation of the inertial position of the standard IKONOS orbit to an intersection with the ray from the scene center in azimuth and elevation direction, the inertial orbit can be determined (Figure 2.8 left). If the GCPs are located in different height levels, it is also possible to determine the satellite orbit without information about satellite elevation and azimuth information (Figure 2.8 left). With a simple bias correction the location of GCPs can be respected. In most cases a bias correction by shift was satisfactory for IKONOS images (Jacobsen, 2002).

An approximate simple orientation method is the 3-D affine transformation (equation 2.1.2.4) (Fraser and Hanley, 2005), being successful for flat areas. The 3-D affine transformation is a parallel projection; this is satisfactory for the orbit direction if the computation is made in national net coordinates but, in line direction, perspective geometry exists causing discrepancies in mountainous areas. An extension to an extended 3-D transformation (equation 2.1.2.5) (Jacobsen, 2007) solved the problem as well as the influence of asynchronous imaging. IKONOS images were available as geo-images, being a projection to a plane in object space. For this image geometry these equations can be used, but not for original images: they require an additional extension to 3-D affine transformation for original images (equation 2.1.2.6) (Jacobsen, 2007).

$$x_{ij} = a1 + a2 * X + a3 * Y + a4 * Z \qquad (2.1.2.4)$$
$$y_{ij} = a5 + a6 * X + a7 * Y + a8 * Z$$

$$x_{ij} = a1 + a2 * X + a3 * Y + a4 * Z + a9 * X * Z + a10 * Y * Z \qquad (2.1.2.5)$$
$$y_{ij} = a5 + a6 * X + a7 * Y + a8 * Z + a11 * X * Z + a12 * Y * Z$$

$$x_{ij} = a1 + a2 * X + a3 * Y + a4 * Z + a9 * X * Z + a10 * Y * Z + a13 * X * X \qquad (2.1.2.6)$$
$$y_{ij} = a5 + a6 * X + a7 * Y + a8 * Z + a11 * X * Z + a12 * Y * Z + a14 * X * Y$$

Another simplified scene orientation is the direct linear transformation (DLT). With 11 unknowns perspective geometry is used, which is not exactly fitting to the satellite scene geometry. It also compensates the inner orientation what is doubtful for the small field of view of satellite images, leading to strong correlations of the unknowns. The DLT cannot be recommended, which is also confirmed by the results given in Figure 2.8 (right).

Figure 2.8 (right) gives an overview of the success of scene orientation with different orientation methods for QuickBird depending upon the number of GCPs used. The results of the geometric reconstruction of the images and the bias-corrected RPC

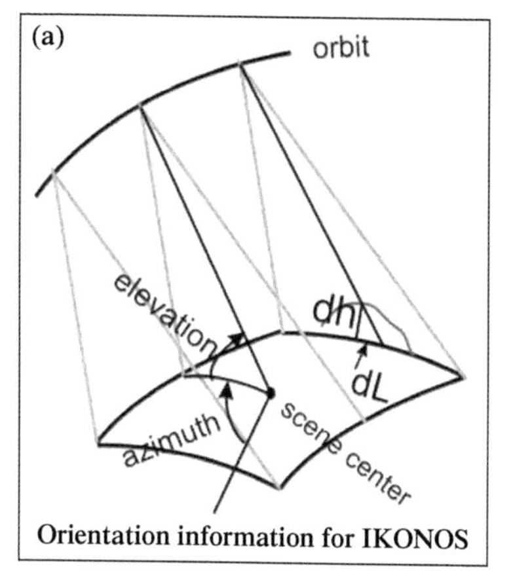

Orientation information for IKONOS

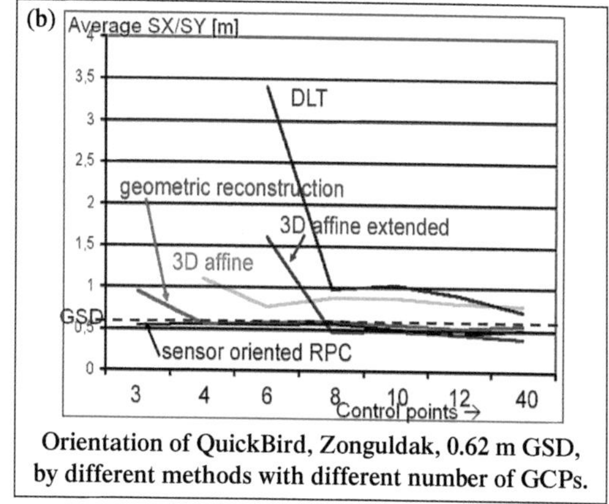

Orientation of QuickBird, Zonguldak, 0.62 m GSD, by different methods with different number of GCPs.

Figure 2.8 Scene orientation with alternative orientation methods. (a) scene orientation approach with alternative methods, (b) accuracy results obtained over Zonguldak test site.

are close together. With only just 3 GCPs the RPC orientation leads to better results. With 8 GCPs and more, the orientation with the extended 3D-affine transformation is approaching the same accuracy. In mountainous areas the DLT is not as good, and the 3D-affine transformation is also limited. This behaviour in orientation methods was confirmed with other experiments. Orientation of original OrbView-3 images (Büyüksalih, Akcin and Jacobsen, 2006) gave approximately 50% better accuracy by RPC orientation than with 3D-affine transformation for the original data; the result based on DLT was poor.

Depending upon the base-to-height relation, object heights are more sensitive to geometric differences than horizontal components, as shown by equation 2.1.2.7.

$$DZ = Dpx * h/b \tag{2.1.2.7}$$

Influence of parallax differences (Dpx) to differences in Z (DZ), where $h$ = flying height above ground; $b$ = base; $x$-parallax differences Dpx = $x' - x''$

$$SZ = h/b * Spx \tag{2.1.2.8}$$

Standard deviation of height (SZ) as a function of the standard deviation of the $x$-parallax (Spx [GSD]), where Parallax = difference of the image components in the direction of flight px = $x' - x''$

The height-to-base relation is usually in the range of 1.6 but, in special cases, it may be quite a bit larger, so the object height is more strongly influenced by $x$-parallax in such a case.

The requirement and influence of the affine parameters for the bias correction depends upon the sensor used. For IKONOS, usually the bias correction by shift parameters (A1 and A4 in equation 2.1.2.3) was satisfactory, but for Cartosat-1 (2.5 m GSD) all affine parameters were required (Jacobsen, 2017). In the test area of Warsaw,

Cartosat-1 images showed root mean square errors at 33 GCPs of 16.86 m and 12.54 m by RPC-orientation with bias correction by shift. They were reduced to 1.41 m and 1.35 m by bias correction with affine transformation, corresponding to approximately 0.55 GSD, which is a good result for the used GCPs. For Pleiades images, only one of the affine parameters A2, A3, A5 and A6 was significant, reducing the root mean square differences in the test area of Zonguldak (Turkey) by approximately 2% at 168 GCPs. Nevertheless the bias correction by affine transformation caused a height model tilt of approximately 2 m over the scene range.

The bias correction by affine transformation of a WorldView-2 (0.5 m GSD) stereo pair gave at 14 GCPs a reduction in the root mean square differences of 24%, to 0.54 m against a bias correction by shift. All affine parameters are significant but this caused a tilt of the height model by 0.5 m over the scene range.

In the ISPRS test area of Sainte-Maxime, for a Ziyuan-3 stereo pair (3.5 m GSD) on average the root mean square difference at 12 GCPs was improved by bias correction with affine transformation by 34% against a bias correction by shift. The average accuracy is 2.03 m and 3.08 m, corresponding to 0.58 GSD and 0.88 GSD, respectively. Furthermore, here a height model tilt of up to 5 m over the whole scene range was caused by the bias correction by affine transformation.

Problems in the image geometry may be analyzed by the *y*-parallaxes of intersections of corresponding rays for the generation of digital surface models (DSM). The high number of intersections for a stereo model – tens of millions – allows the elimination of random errors due to the high number, so systematic *y*-parallaxes can even reveal small problems of the image geometry.

The *y*-parallaxes of a height-model computation with WorldView-2 images have been averaged in 200 equidistance groups as a function of *X* and *Y* (Figure 2.9). In the area Karaburun, Turkey, three WorldView-2 stereo models, located beside each other, have been taken from the same orbit. All three models show periodic errors , especially in *X*-direction. The period of the discrepancies corresponds to the 64 panchromatic CCD-lines, arranged by 32 elements in two directly neighboring locations of the focal plane. The deviation from a straight line in the *Y*-direction may be explained by satellite jitter. The linear difference of the systematic *y*-parallax errors (Figure 2.9,

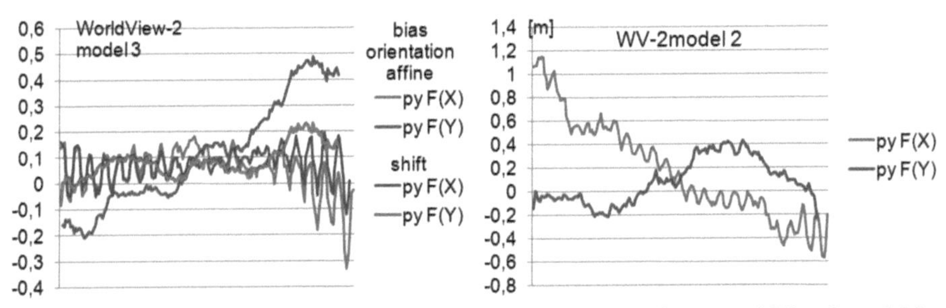


**Systematic *y*-parallaxes, WorldView-2, model 3, bias correction by shift and affine transformation.**

**Systematic *y*-parallaxes, WorldView-2, model 2, bias correction by affine transformation.**

Figure 2.9 Systematic *y*-parallaxes as function of *X* and *Y*.

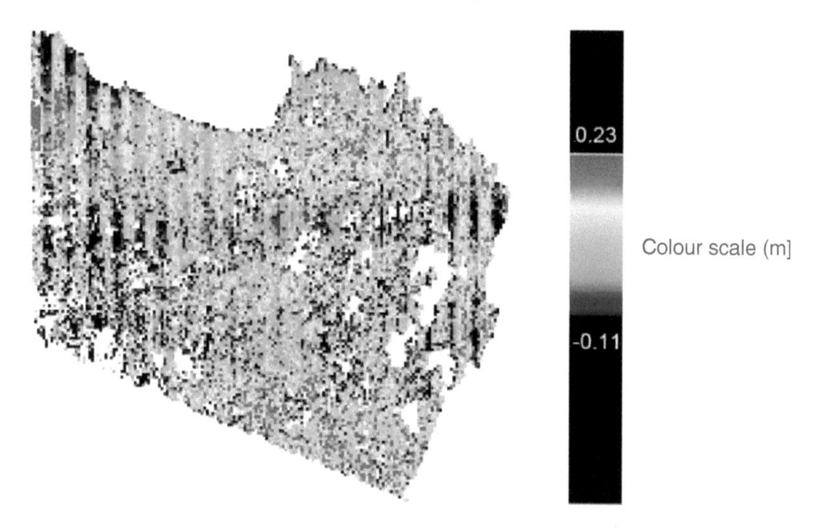


Figure 2.10 Systematic $y$-parallaxes of WorldView-2, model 3, distributed over the stereo model.

left) are caused by the bias correction with affine transformation. In model 2 (Figure 2.9, right) a tilt in the $X$-direction of approximately 1.4 m over the whole scene is caused by the same reason.

Figure 2.10 shows the systematic $y$-parallaxes of the same model 3 in Karaburun as a function of location in the stereo model. The periodic errors as a function of $X$ are visible by the striping. Gaps in the model are caused by water surfaces.

The systematic $y$-parallaxes of Ziyuan-3 (3.5 m GSD) in Sainte-Maxime (Figure 2.11) are strongly influenced by the affine bias correction with a tilt in the $Y$-direction of approximately 3 m. Independent upon this trend, in the $Y$-direction the limited accuracy of the attitude can be seen and in the $X$-direction sub-optimal sensor calibration is observed (Zhang *et al.*, 2015).

Cartosat-1 (2.5 m GSD) shows, in the test areas of Warsaw and Mausanne, periodic errors in the $Y$-direction with amplitude of approximately 0.4 m, which may be caused

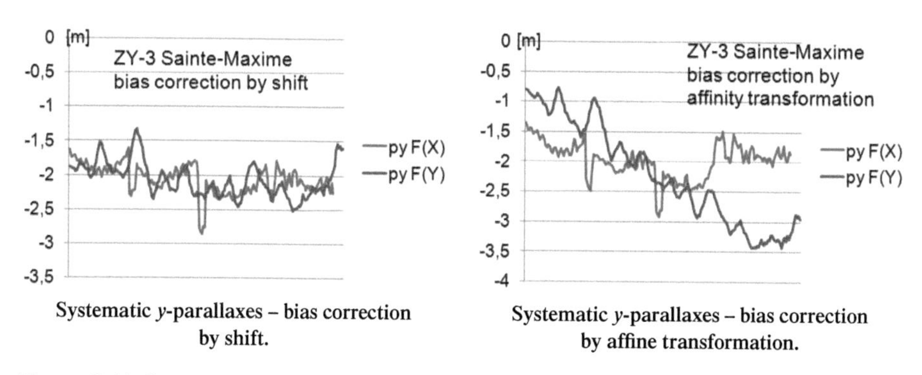


**Systematic $y$-parallaxes – bias correction by shift.**

**Systematic $y$-parallaxes – bias correction by affine transformation.**

Figure 2.11 Systematic $y$-parallaxes of Ziyuan-3 (ZY-3).

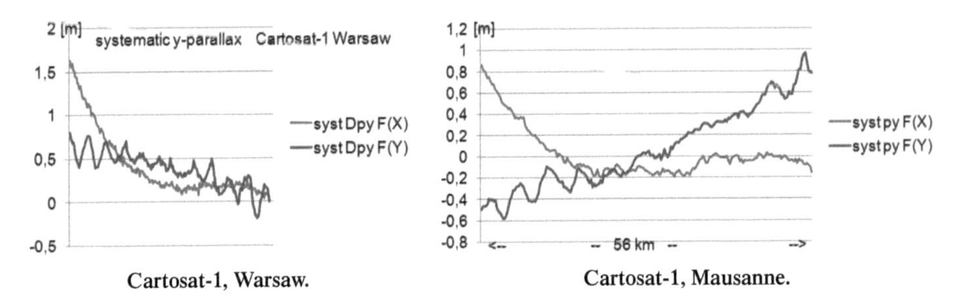


Figure 2.12 Systematic *y*-parallaxes of Cartosat-1.

by satellite jitter, and in the *X*-direction a curvature of approximately 0.5 m occurs (Figure 2.12). Such deformations are usually ignored due to the limited size in relation to the GSD. The required bias correction by affine transformation in the test area of Mausanne leads to a tilt in the *Y*-direction.

In the project area of Riyadh, IKONOS (1 m GSD) has a very linear trend in the *X*-direction with maximal deviations up to 10 cm, while in the *Y*-direction systematic errors up to 30 cm appear, caused by the attitudes. In the same area GeoEye (0.5 m GSD) shows differences against linearity of 10 cm in the *X*- and *Y*-direction.

One exception is Pleiades (0.5 m GSD). In the test area of Zonguldak the systematic *y*-parallaxes are limited to approximately 4 cm, which is just 8% of the GSD. The differences against linearity of the systematic *y*-parallaxes are approximately 2 cm, and in one location 3cm.

The *y*-parallaxes have an influence of only half the size to the ground coordinates, but they indicate corresponding problems for the *x*-parallax, influencing the object height in a stereo model. The accuracy of the horizontal ground coordinates is indicated by the discrepancies at the ground control points. With satisfactory object point identification, under operational conditions a standard deviation of the ground coordinates in the *X*-direction (SX) and in the *Y*-direction (SY) of approximately 0.6 GSD can be reached by bias-corrected RPC orientation or by geometric reconstruction using the orbit information. The dominating influence on the accuracy is the identification of the object points. Fraser and Hanley (2005) used centers of roundabouts and circular buildings as GCPs and independent check points for the orientation and validation of IKONOS and QuickBird images. Such an optimal object point definition leads to SX and SY of approximately 0.25 GSD, but such object points are very rare and usually not available.

Under operational conditions the greatest accuracy is required for the determination of height models, due to the dependency of the ground height on the height-to-base relation (equation 2.1.2.7) and the dependency upon usually two images. Problems of image orientation are shown in Table 2.7 for the stereo satellite Ziyuan-3, which has a limited accuracy for the attitude determination and calibration (Zhang *et al.*, 2015). The ISPRS test data set for Sainte-Maxime has been used for the analysis of Ziyuan-3. For this test data set, GCPs and a reference surface model from the French IGN are available. The reference height model based on aerial images has an SZ of

**Table 2.7** Comparison of Ziyuan-3 DSM in non-forest areas with reference DSM from aerial images, depending upon type of bias correction for RPC orientation.

| | Whole non-forest area | | Slope < 10% | | Tilt over scene range | |
|---|---|---|---|---|---|---|
| Sensor orientation and post-correction | SZ [GSD] | NMAD [GSD] | SZ [GSD] | NMAD [GSD] | X [GSD] | Y [GSD] |
| 2D shift | 1.11 | 0.82 | 0.78 | 0.67 | –0.65 | –0.98 |
| 2D affine | 1.33 | 1.20 | 1.02 | 1.02 | 0.31 | –3.32 |
| 3D shift | 1.11 | 0.84 | 0.78 | 0.69 | –0.45 | –1.17 |
| 3D affine | 1.22 | 1.03 | 0.91 | 0.89 | 0.70 | –2.20 |
| 3D shift, leveled | 1.10 | 0.80 | 0.76 | 0.54 | | |
| 3D affine, leveled | 1.10 | 0.79 | 0.78 | 0.54 | | |

3D = three-dimensional post-correction of DSM. NMAD = normalized median absolute deviation.

approximately 20 cm. For the determination of the Ziyuan-3 DSM, both inclined views with 3.5 m GSD and not the nadir image with 2.1 m GSD were used, leading to a stereo model with a base-to-height relation of 1:1.15. This is in accordance with an unusual large angle of convergence. The corresponding image positions have been determined by least squares matching with the region growing due to the dominating rural area. Because of DSM definition problems in the forest, the analysis has been done only for the non-forest area (Table 2.7) to be able to determine the system accuracy. For the results in Table 2.7 the dimension of GSD is used to allow a better comparison with other space data. The scene orientation has been made with bias-corrected RPCs (see above), by bias correction with shift and with affine transformation. The problem of model tilt by bias correction with affine transformation has been mentioned above, so an attempt was made to improve this situation by employing post-correction of the DSM by shift, affine transformation and rotation around *X*- and *Y*-axis.

The accuracy figures are shown for the whole non-forest area and separately for the areas with terrain slope below 10%, due to the dependency of the DSM accuracy on the terrain slope. In addition to the standard deviation, the normalized median absolute deviation (NMAD) is shown (Höhle and Höhle, 2009). NMAD is based on the median absolute differences multiplied by the factor 1.4826. In the case of normally distributed differences, NMAD for the height is identical to SZ, but usually there are some deviations from normal distribution due to unequal conditions for the whole data set as caused by the influence of terrain slope or by a smaller number of larger discrepancies, which may be caused by buildings and trees and by limitations of the mask for the specification of non-forest area.

Figure 2.13 shows two typical examples of the frequency distribution of height differences overlaid with normal distribution based on SZ and NMAD. The left-hand side corresponds to Ziyuan-3 DSM against the reference DSM for terrain inclination below 10%, based on 2.7 million differences. Due to the high number, the frequency distribution is smooth. On the right-hand side the frequency distribution of height differences of a Pleiades DSM against a reference DSM with just 3,752 check points is shown – with such a limited number the frequency distribution is not smooth.

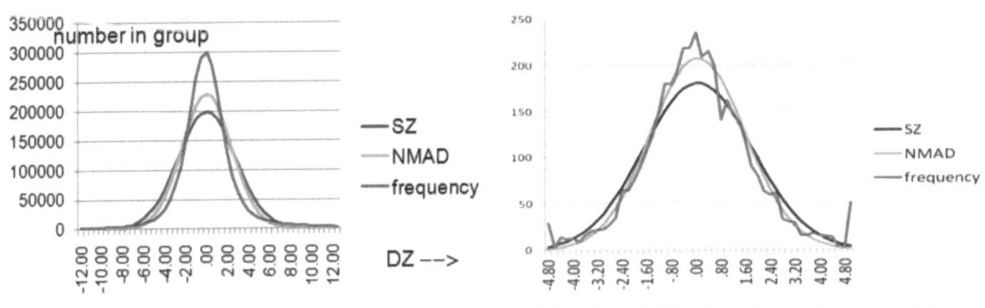


**Ziyuan-3 DSM against reference DSM.** **Pleiades-1A DSM against reference points.**

Figure 2.13 Frequency distribution of height differences overlaid with normal distribution based on SZ and NMAD.

As is typical, the normal distribution based on NMAD fits better to the frequency distribution than the normal distribution based on SZ. That means NMAD describes the frequency distribution better than SZ.

Depending upon the orientation used (Table 2.7), the results of the Ziyuan-3 height models show very clearly the problems of bias correction by affine transformation. For Ziyuan-3, not only is the bias correction by affine transformation tilting the model but the bias correction may also lead to a not negligible height model tilt, which can only be explained by the reported limited attitude accuracy. A three-dimensional orientation slightly improves the model tilt. If the height model is leveled based on a reference DSM (Figure 2.14, right), the type of orientation is unimportant – even with separate orientations of the scenes (2D orientation), after leveling the same result has been reached. Under operational conditions no precise reference height model is available, but a similar improvement was possible also with the Shuttle Radar Topography Mission Digital Surface Model (SRTM DSM) and AW3D30, which are available free of charge.

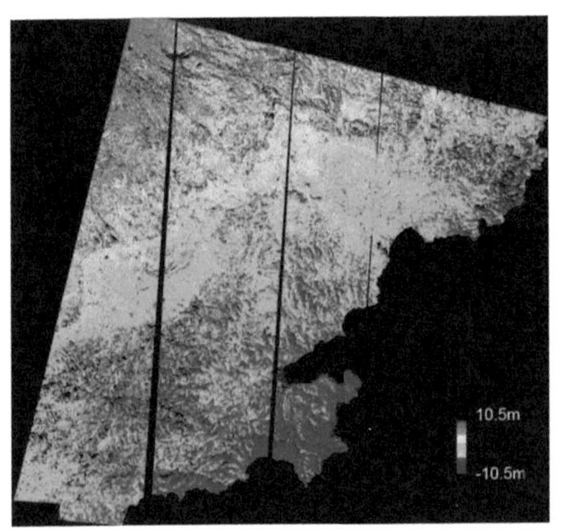


**Colour-coded height differences, whole area.** **Colour-coded height differences after leveling.**

Figure 2.14 Ziyuan-3 DSM based on bias correction by affinity transformation.

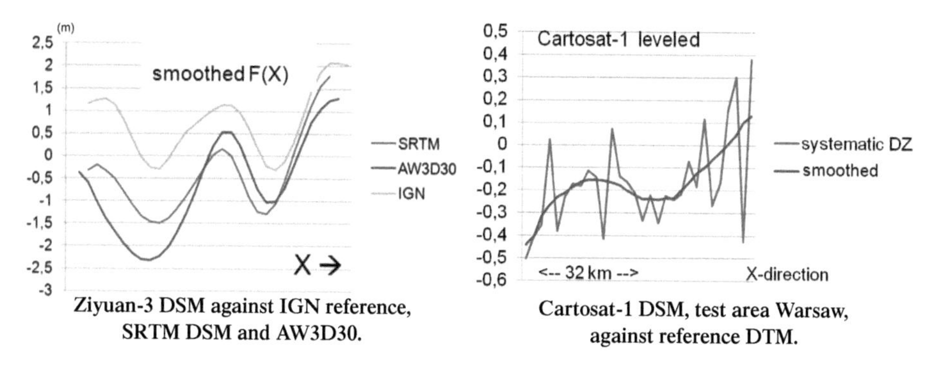


Figure 2.15 Smoothened systematic height differences as a function of *X*.

Apart from model tilt, an undulation in the *X*-direction with amplitude of +/- 1 m may be observed (Figure 2.15, left). Other space image models also display undulations in the *Y*-direction. Even if the undulation of the Ziyuan-3 DSM is small against the GSD of 3.5 m and the accuracy numbers are not so much influenced by this, it should not be neglected and needs to be corrected. Similar systematic height differences also exist in other height models from satellite images (Jacobsen, 2016, 2017). The systematic height differences have to be filtered, as shown in Figure 2.15 (right); this is required due to changed vegetation height or different interpretations of the visible surface as a function of the ground resolution.

For other methods of image matching, epipolar images are required. Epipolar lines are defined by the plane containing the projection centers of both images of a stereo model and a ground point, intersecting the image plane. By the definition of space images as line scanner images, only quasi-epipolar images can be generated, where corresponding image points have the same *y*-coordinates, reducing the search of corresponding points to the *x*-direction in the epipolar image. The generation of quasi-epipolar images depends upon the type of space images. Images projected onto a plane with constant height, as IKONOS Geo processing level, only have to be rotated to the base direction. The remaining *y*-parallaxes are in the sub-pixel range. The base direction can be computed from information in the metadata available for all space images. Original images, mostly available for stereo pairs, at first have to be transformed to images projected onto a plane with constant height. This need not necessarily be done in a tangential coordinate system; it can also be done in a national coordinate system, e.g. UTM.

In non-forest areas with terrain inclination below 10%, for all satellite image types using the correct orientation and post-processing a standard deviation of height of approximately 0.6 GSD has been attained with stereo pairs having a base-to-height relation of 1:2 and better. The main problem is not so much the sensor geometry as the detailed description of the surface.

### 2.1.3 Satellite images for 3D city modelling

The generation of height models for predominantly rural areas has been described above. For built-up areas and rough mountains, we have the problem that the height

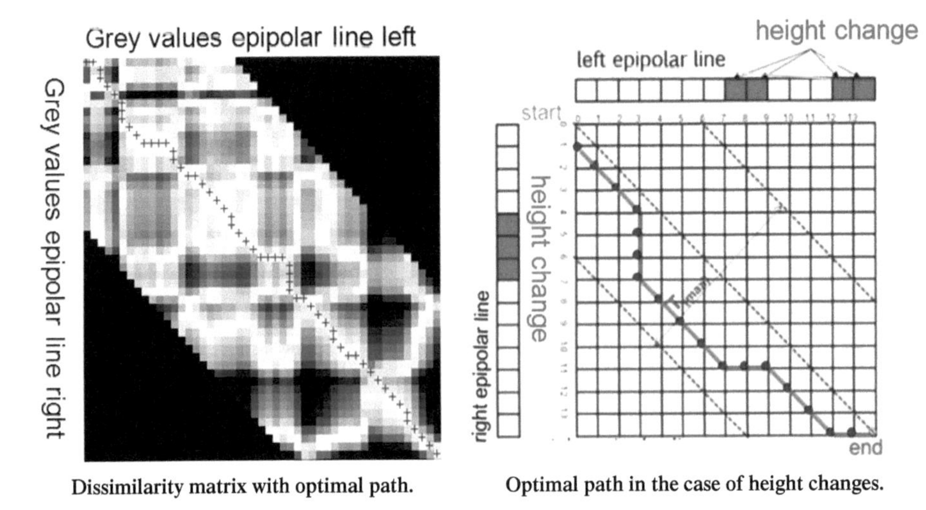


**Dissimilarity matrix with optimal path.** **Optimal path in the case of height changes.**

Figure 2.16 Dissimilarity matrix used by matching with dynamic programming.

does not change continuously. Of course, this could be acknowledged through the use of area-based matching with manually measured break lines, but this is very time consuming. For urban areas and rough mountains, the matching should be done by semi-global matching (SGM) (Hirschmüller, 2005), which does not smooth out the determined object surface, representing the building structure at pixel level.

Some program packages are combining SGM with area-based matching, switching from one method to the other depending upon the area type. SGM is an improvement of the matching by dynamic programming (DP) (Birchfield and Tomasi, 1998).

In DP corresponding epipolar lines are combined to form a dissimilarity matrix, showing the grey value differences of the normalized epipolar lines (Figure 2.16). If the epipolar line crosses a building, the object height goes up suddenly and the *x*-parallax is changing, leading to an optimal path as shown in Figure 2.16 (right). The calculated optimal path shown in Figure 2.16 (left) corresponds to the optimal cost function (Birchfield and Tomasi, 1998). The DP has the disadvantage that neighboring epipolar lines are handled independent from each other, leading to striping of the generated height model, which can be reduced by filtering. SGM improves DP by using not only the epipolar lines but also lines in several directions in relation to the handled object point.

Figure 2.17 demonstrates the smoothing of buildings using area-based matching. Of course, the results based on a small template of 5 × 5 pixels is better than for a larger template, but with a small template the results are not very reliable, leading to errors or unacceptable values. With pixel-based matching by DP or SGM the shape of the building at the street level does not exactly match to the shape of the building at the top (Figure 2.17). It depends on the fact, that we usually only have a 2.5D database that has only one height value per pixel and the occlusions as a function of the view direction.

The occlusion is increased by the fact that the satellite usually does not pass directly over the project area. In most cases, the image order conditions allow an incidence angle of up to 20° for viewing to the left or right of the orbit.

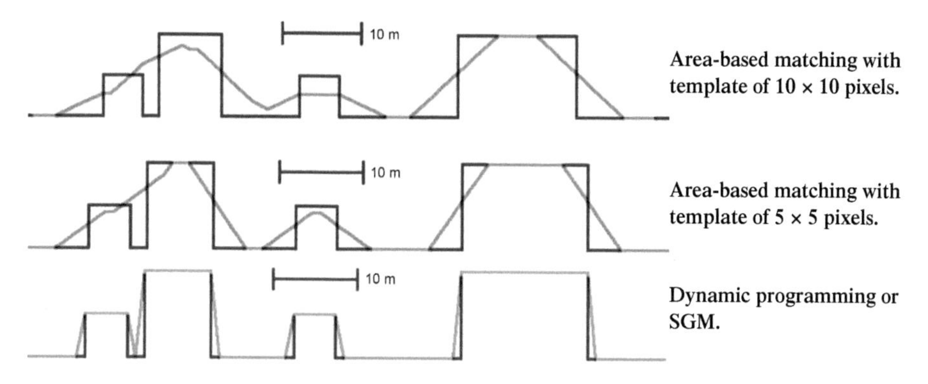


Figure 2.17 Simulated height profile (green line) based on area- and pixel-based matching with images of 1 m GSD and nadir angles of +17° and –17° in a built-up area.

As shown in Figure 2.19, taken over Istanbul, in the case of a WorldView-4 stereo model for the standard height of old living buildings, the street width across orbit direction must be at least 10.5 m for the existing imaging configuration shown in Figure 2.18 for stereo coverage to reach the street level. In addition, few pixels with satisfying contrast on the street are required to determine the height of the street. In old residential areas in Istanbul there is often no such street width. Experience has shown that the angle of convergence in built-up areas should not exceed 20°.

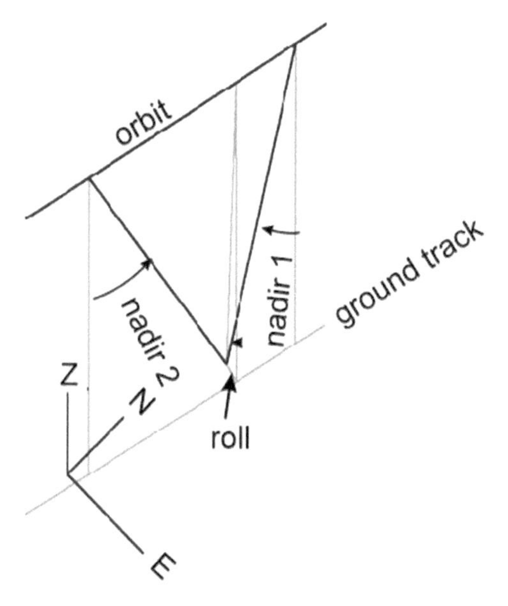


Figure 2.18 View configuration of a WorldView-4 stereo model.

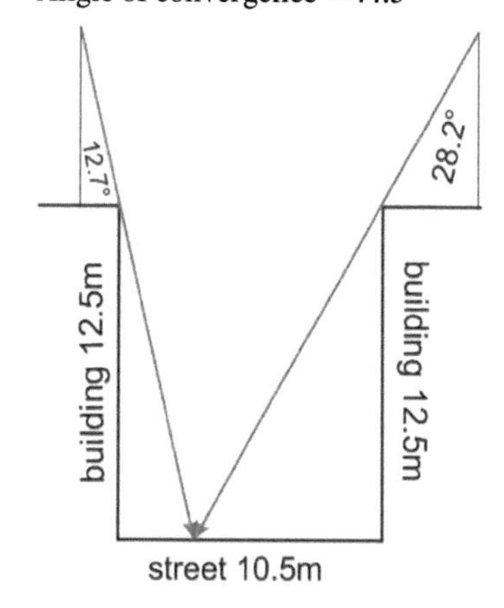


Figure 2.19 Occlusion by buildings.

**Least squares matching** **DP with median filter** **SGM**

Figure 2.20 Height model using IKONOS stereo model with different matching methods.

The simple formula (equation 2.1.2.8) for the accuracy of the object height shows a higher accuracy for a larger angle of convergence, which corresponds to the base-to-height relation, but for a smaller angle of convergence both images of a stereo model are more similar, leading to matching to a smaller value of the standard deviation of the *x*-parallax, Spx. So, in built-up areas the height accuracy does not depend so much upon the angle of convergence. Experience has shown that the height accuracy in built-up areas is improved if the angle of convergence exceeds 20°.

As Figure 2.20 shows (Alobeid, Jacobsen and Heipke, 2009), by least squares matching, the building shapes are not as clear and the facades are inclined. Even with a median filter of 1 × 7 pixels across the epipolar line the results of DP achieved for the roads are not optimal. With SGM the most accurate result was reached, showing the object shapes better than with the other methods. The rough DSM shows a park with trees in the lower left corner.

Figure 2.21, based on a Kompsat-3 stereo model in a densely built-up area of Istanbul (Büyüksalih, Bayburt and Jacobsen, 2018), shows the limitations of SGM. Kompsat-3 has a GSD of 0.71 m. The angle of convergence at 44.7° is larger than suggested above, causing problems in separating neighboring buildings due to occlusions. The black lines around the buildings are also caused by the larger incidence angles. The larger black areas are partially caused by trees, but also by dark shadows next to higher buildings. For a 3D city model, therefore, post-processing is required. Nevertheless, there are enough points on the bare ground to filter the DSM to a DTM and to generate an normalized Digital Surface Model (nDSM) with the building height above ground by the difference of both.

The visible surface is determined by image matching. Often, digital terrain models (DTM) with the height of the bare ground are required. If at least some points of the bare ground are included in the DSM then, by filtering, a DTM can be generated (Passini, Betzner and Jacobsen, 2002). In closed forest areas the filtering has limited success. Here, a correction by radar profile information as used by Smith and Berry (2008) for the correction of the SRTM DSM to a DTM with the name ACE-2 may help. Nevertheless ACE-2 is not corrected optimally.

According to the image resolution SGM leads to optimal description of the building shape. Nevertheless, SGM also has some problems in densely built-up areas. The height determination in dark shadows may be difficult, which may be improved by Wallis filtering (Büyüksalih *et al.*, 2009). In addition, for very high buildings the

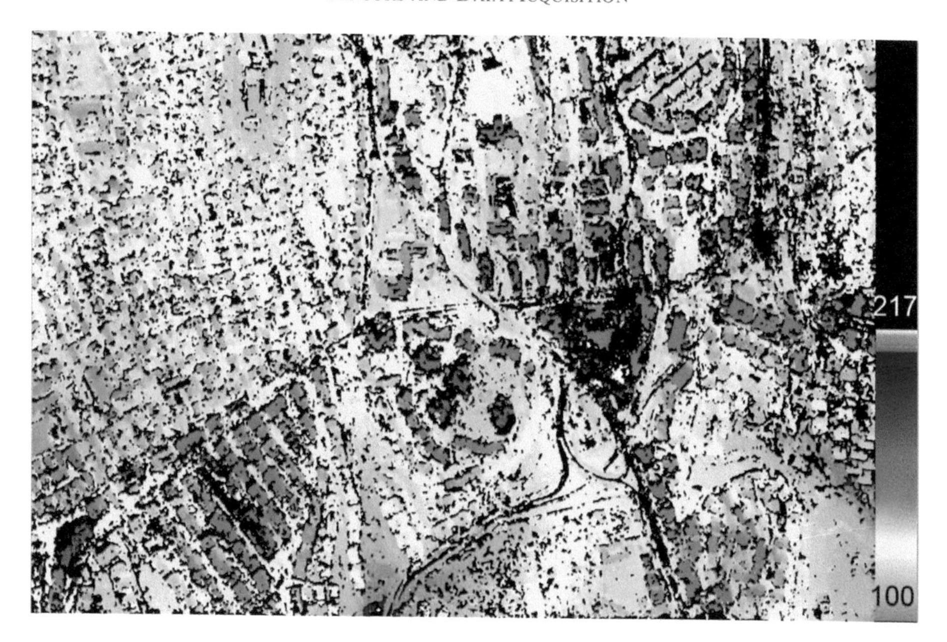


Figure 2.21 Colour-coded height model by SGM from Kompsat-3; width of DSM = 1 km.

search range in the dissimilarity matrix (Figure 2.16) is sometimes exceeded, especially for larger angles of convergence, resulting in gaps in height determination. In addition, some noise occurs at building edges, which requires post-processing.

### 2.1.4 Products generated by satellite images

Height models, as described above, are a typical product generated by satellite stereo or tri-stereo models. The products most often used are orthoimages. For a grid specified in object space the grey values of the used bands in the image positions corresponding to the ground coordinates are used. The transformation of the object points into the images will not fall exactly into the pixel centers, requiring an interpolation. Without interpolation the nearest neighborhood leads to images with grey value staircases when zoomed in detail, so most often a bi-linear interpolation of the grey values is applied. A cubic convolution is also in use, but it includes some filter effects.

The geometric quality of orthoimages mainly depends on the quality of the height model used: this should be accurate enough in relation to the GSD and the incidence angle (DP = DZ * tan $i$; DP = error in position; DZ = height error; $i$ = incidence angle). Under incidence angle we understand the nadir angle from the ground to the satellite, which is not the same as the nadir angle from the satellite to the ground, due to earth curvature. A basic problem of orthoimages is caused by sudden object height changes, as occurs with buildings or bridges. In case of an inclined view direction we are able to see building facades, but in an orthoimage the roof should fit to the basement and building leaning effects should be minimal. Bridges over a valley, where the height model describes the valley, are deformed. Elements directly located behind a building

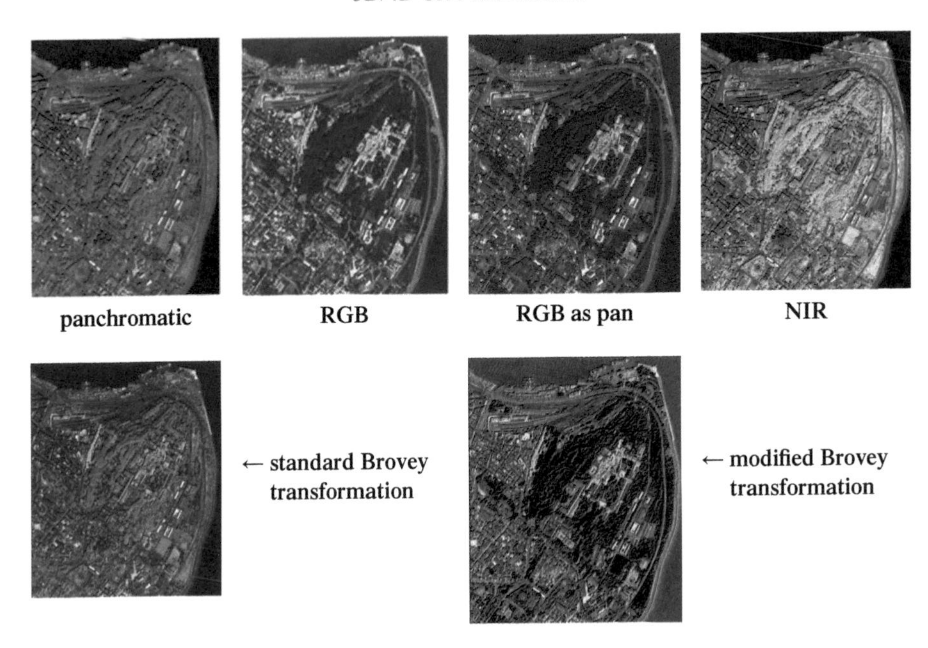


Figure 2.22 Pan-sharpening of IKONOS image.

are occluded. Only with images taken from different directions and an accurate DSM, fitting as much as possible to the 3D shapes, can true orthoimages be generated free from such effects.

Orthoimages should usually be based on pan-sharpened images, but they should also correspond to our expectation based on the sensitivity of the human eye. If the spectral range of the panchromatic sensor does not correspond to the visible range, as is the case for IKONOS, a simple pan-sharpening by Brovey transformation (Figure 2.22) leads to strange colours (Figure 2.22 lower left). With a modified Brovey transformation, where parts of the NIR influence is subtracted from the panchromatic channel, an image corresponding to the expectation of the user can be generated.

$$DN_{fused} = \frac{DN_b * 3}{DN_b + DN_g + DN_r} * DN_{pan} * MF \qquad (2.1.4.1)$$

Standard Brovey transformation where $DN_b / DN_g / DN_r / DN_{NIR} / DN_{pan}$ = grey value of blue/green/red/near infrared/panchromatic band; $MF$ = scale factor for grey values – may be different for R, G, B

$$DN_{fused} = \frac{DN_b * 3}{DN_b + DN_g + DN_r} * \frac{DN_{pan} - F * \frac{DN_{nir}}{3}}{1 - F/3} * MF \qquad (2.1.4.2)$$

Modified Brovey transformation where $F$ = factor for respecting NIR.

Orthoimages are required to correspond to the resolution of the human eyes at usual reading distance, which is 8 pixels/mm (1 pixel/0.125 mm), in order to give us the impression of smooth images. So scale numbers with GSD/0.125 mm may be accepted, corresponding to a scale 1:4000 for 50 cm GSD (scale number = 500 mm/0.125 mm).

Satellite images can be also used for manual generation of line maps. This is possible in a stereo model or by on-screen digitizing of orthoimages. The three-dimensional view has advantages for object identification but requires better trained operators. In some tests comparing both methods, only very few differences between the methods were apparent. The colour is more important – without colour information some interpretation errors have occurred.

Experience with the generation of topographic maps based on high- and very high resolution space images (Jacobsen and Büyüksalih, 2006; Topan *et al.*, 2009) has resulted in the requirement of 0.1 mm GSD at a map scale of 0.1 mm × 50000, 5 m, GSD for the map scale 1:50 000, or 50 cm GSD maps at the scale 1:5 000. In the case of large and clear objects even 0.05 mm in the map scale can be used, corresponding to a scale 1:2500 with 0.5 m GSD. But the GSD should not exceed 5 m, due to some important elements being independent of the scale. The scale for thematic maps depends upon the required objects – this may be different to topographic maps.

## References


Alobeid, A., Jacobsen, K. and Heipke, C. (2009) Building height estimation in urban areas from very high resolution satellite stereo images. ISPRS Hannover Workshop 2009, *IAPRS*. Vol. XXXVIII-1-4-7/W5.

Birchfield, S. and Tomasi, C. (1998) A pixel dissimilarity measure that is insensitive to image sampling. *IEEE Transactions on Pattern Analysis and Machine Intelligence*, 20(4), 401–406.

Büyüksalih, G., Akcin, H. and Jacobsen, K. (2006) Geometry of OrbView-3 images. ISPRS Ankara Workshop 2006: Topographic Mapping from Space. WG I/5; I/6, IAPRS XXXVI-1/W41.

Büyüksalih, G., Bayburt, S. and Jacobsen, K. (2018) Analysis of height models based on KOMPSAT-3 images. Gİ4DM 2018: Geoinformation for Disaster Management. ISPRS Archives XLII-3-W4-115-2018.

Büyüksalih, G., Baz, I., Bayburt, S., Jacobsen, K. and Alkan, M. (2009) Geometric mapping potential of WorldView-1 images. ISPRS Hannover Workshop 2009. *IntArchPhRS*. Vol. XXXVIII-1-4-7/W5.

Büyüksalih, G. and Jacobsen, K. (2005) Optimized geometric handling of high resolution space images. ASPRS annual convention, Baltimore.

d'Angelo, P., Kuschk, G. and Reinartz, P. (2014) Evaluation of Skybox video and still image products. IAPRS XL-1.

Doyle, F. J. (1996) Thirty years of mapping from space. *IAPRS*, Vol XXXI, Part B4, Vienna.

Fraser, C. and Hanley, H. (2005) Bias-compensated RPCs for sensor orientation of high-resolution satellite imagery. *PERS*, 2005, 909–915.

Grodecki, J. (2001) IKONOS stereo feature extraction – RPC approach. ASPRS Annual Conference, St Louis, MO.

Grodecki, J. and Dial, G. (2003) Block adjustment of high-resolution satellite images described by rational polynomials. *Photogrammetric Engineering and Remote Sensing*, 69(1), 59–68.

Hirschmüller, H. (2005) Accurate and efficient stereo processing by semi-global matching and mutual information. *IEEE Conference on Computer Vision and Pattern Recognition CVPR '05*, Vol. 2, San Diego, CA, 807–814.

Höhle, J. and Höhle, M. (2009) Accuracy assessment of digital elevation models by means of robust statistical methods. *ISPRS Journal of Photogrammetry and Remote Sensing*, 64, 398–406.

Jacobsen, K. (1997) Calibration of IRS-1C PAN-camera. Joint Workshop: Sensors and Mapping from Space, Hannover 1997. http://www.ipi.uni-hannover.de/ [accessed May 2017].

Jacobsen, K. (2002) Mapping with IKONOS images In T. Benes (ed.), *Geoinformation for European-wide Integration* (pp. 149–156). Millpress, Rotterdam.

Jacobsen, K. (2006) Calibration of imaging satellite sensors. ISPRS Ankara Workshop 2006 Topographic Mapping from Space, WG I/5; I/6, *IAPRS*, Vol. XXXVI-1/W41.

Jacobsen, K. (2007) Orientation of high-resolution optical space images. ASPRS annual conference, Tampa, FL.

Jacobsen, K. (2011) Characteristics of very high resolution optical satellites for topographic mapping. ISPRS Hannover Workshop 2011, *IAPRS* Vol. XXXVIII-4/W19.

Jacobsen, K. (2016) Analysis and correction of systematic height model errors. *IAPRS* Vol. XLI-B1, 333–339.

Jacobsen, K. (2017) Problems and limitations of satellite image orientation for determination of height models. ISPRS Hannover Workshop, *IAPRS*, Vol. XLII-2/W5.

Jacobsen, K. and Büyüksalih, G. (2006) Mapping from space – A Cooperation of Zonguldak Karaelmas University and University of Hannover. Conference: 5th Turkish-German Geodetic Days, Berlin, 28–31 March 2006.

Jacobsen, K., Topan, H., Cam, A., Özendi, M. and Oruc, M. (2016) Image quality assessment of Pleiades-1A Triplet Bundle and pan-sharpened images, *PFG* 3/2016, 75–86.

Kocaman, S. and Gruen, A. (2007) Orientation and calibration of ALOS/PRISM imagery. *IAPRS*, Vol. XXXVI/1-W51.

Passini, R., Betzner, D. and Jacobsen, K. (2002) Filtering of digital elevation models, ASPRS annual convention, Washington, DC.

Schneider, T., Seitz, R., Förster, B. and Jacobsen, K. (2001) Remote sensing based parameter extraction for erosion control purposes in the loess plateau of China: High resolution mapping from space 2001. Hannover. https://www.ipi.uni-hannover.de [accessed May 2017].

Schroeder, M., Müller, P. and Reinartz, P. (1999) Vicarious radiometric calibration of MOMS at La Crau test site and intercalibration with SPOT. Workshop on Sensors and Mapping from Space, Hannover, Germany.

Smith, R.G. and Berry, P.A.M. (2008) ACE2 New global digital elevation model: Case studies of rainforest & dunes. IAG International Symposium on Gravity, Geoid & Earth Observation 2008, Chania, Crete, 23–27th June.

Topan, H., Oruc, M. and Jacobsen, K. (2009) Potential of manual and automatic feature extraction from high-resolution space images in mountainous urban areas. ISPRS Hannover Workshop 2009, IAPRS. Vol. XXXVIII-1-4-7/W5.

Valorge, C., Meygret, A., Lebègue, L., Henry, P., Bouillon, A., Gachet, R., Breton, E., Léger, D. and Viallefont, F. (2004) Forty years of experience with SPOT in-flight calibration. In S. Morain and A. Budge (eds) *Post-launch Calibration of Satellite Sensors* (pp. 119–133). Routledge, London.

Zhang, G. *et al.* (2015) Geometric accuracy improvement and assessment of Chinese high-resolution optical satellites. *IEEE JSTARS*, 2015. 2429151.

# Chapter 3
# Geometric processing for image-based 3D object modelling

*Rongjun Qin and Xu Huang*

## 3.1 Introduction

Image-based 3D object modelling refers to the process of converting raw optical images to 3D digital representations of the objects. Very often, such models are desired to be dimensionally true, semantically labelled, and with a photorealistic appearance (reality-based modelling). Laser scanning was deemed the standard (and direct) way to obtain highly accurate 3D measurements of objects, while one would have to accept the high acquisition cost and its unavailability on some of the platforms. Nowadays, image-based methods supported by recently developed advanced dense image matching algorithms and geo-referencing paradigms, are becoming the dominant approach, mainly due to their high flexibility, availability and low cost. The largely automated geometric processing of images in a 3D object reconstruction workflow, from ordered/unordered raw imagery to textured meshes, is becoming a key part of reality-based 3D modelling. This chapter provides an overview of geometric processing workflow, with a focus on introducing the state-of-the-art methods of three major components of geometric processing: (1) geo-referencing; (2) dense image matching; and (3) texture mapping. Finally, we will draw conclusions and share our outlook of the topics discussed in this chapter.

### 3.1.1 Background

Creating realistic 3D digital models of the physical objects and environment – termed reality-based 3D modelling – is a fundamental research theme. The challenges of achieving such realistic 3D models has, so far, placed a restraint on the increasing demand of different applications, such as CFD (computation fluid dynamics) (Vernay *et al.*, 2015), 3D Geo-database (Haala and Kada, 2010), GIS (geographic information system), urban planning, solar energy potential analysis (Freitas *et al.*, 2015; Strzalka *et al.*, 2011), smart cities etc. (Biljecki *et al.*, 2015; Gruen, 2013). The reconstruction of such models involves a high-level data understanding problem, which it is difficult to reasonably address with state-of-the-art computer algorithms. Existing approaches for generating city-scale and fine-grained 3D models require heavy manual involvement for measuring 3D surfaces, reconstructing topology (e.g. CAD models) and labeling the semantics of the objects (Flamanc *et al.*, 2003). Reality-based modelling (RBM) is a complex task that often requires the success of a chain of important computational solutions, ranging from accurate geometric processing to high-level image and scene understanding (Gruen, 2008; Gruen *et al.*, 2009).

Images/photographs have been the one of the predominant tools for mapping and modelling (paralleling with LiDAR, developed since 1970s (Schwarz, 2010)). LiDAR provides direct and accurate 3D measurements of objects, and has been the standard approach for obtaining highly accurate 3D measurements of objects for the past decades. Recently, the development of automated geometric processing approaches has made images again a much-favored source for 3D modelling (Remondino *et al.*, 2014), since it is a widely available resource on almost all platforms, ranging from personal mobile phones to satellites. 3D modelling from images is no longer expert specific: mobile applications are readily available for turning cellphone images into 3D mesh models; the boost in drones with cameras has made it a low-cost and flexible source for object modelling at fine scales. Software packages are available offering functions that can turn a set of images to 3D mesh models (Agisoft, 2017; Pix4D, 2017; Wu, 2014) with just a few button-clicks. Image-based 3D point clouds are nowadays sufficient for many modelling applications (Gehrke *et al.*, 2010; Nex and Remondino, 2014), and relevant algorithms are consistently and actively improved. In addition, images offer richer textural, spectral and boundary (Huang, 2013) information that is worthy of more scientific exploration in aiding high-level image/data understanding and object labeling.

In general, reality-based modelling (RBM) involves two sets of broadly defined problems: (1) geometric processing, and (2) object labeling and topology reconstruction. Geometric processing (GP) refers to the process of converting raw sensory data all the way to explicit 3D information – e.g. 3D measurements/3D triangle meshes with photorealistic textures. Object labeling and topological reconstruction (Cornelis *et al.*, 2008; Diakité *et al.*, 2014; Liebelt and Schmid, 2010; Verdie *et al.*, 2015), refers to the process of identifying the types of objects and their individual components, as well as the topological relationship between different components of the object.

The geometric processing of images involves image orientation/bundle adjustment, dense image matching, point cloud meshing and texture mapping etc.(Remondino *et al.*, 2014) The second problem set – object labeling and topological identification – is usually operated on sparse/dense point clouds, digital surface models or triangle meshes with textures, and the algorithms are rather independent of datasets, although they may vary slightly with the noise levels and the special character of different types of sensory data.

### 3.1.2 Scope of this chapter

The topic of this chapter is image-based geometric processing in the context of the reality-based 3D modelling. Other modelling techniques that do not rely on raw data, e.g. artificial 3D objects modelling (e.g. for gaming) using rule-based methods, or procedural methods solely based on shape grammars, are considered to be out of this context. RBM is a highly disparate problem composed of many different components in both geometric aspects and high-level image and data understanding, with solutions varying with the types and quality of data. Rather than encompassing all aspects of RBM, this chapter will provide an overview on necessary techniques related to the geometric processing of images for RBM, one of the most relevant and progressive aspects in RBM in recent years. The chapter will focus on three main topics: (1) image geo-referencing; (2) dense image matching, and (3) texture mapping. The image/data understanding part of RBM will not be covered, whereas relevant techniques will be partially included in approaches that incorporate image/data understanding methods for geometric processing. Modelling individual objects and identifying types of objects are addressed in other chapters of this book. In order to outline some of the special characteristics of image-based methods, we will partially include some technical details of data processing techniques for other types of data (e.g. LiDAR). We aim to provide an overview of the complete chain of geometric processing and the main works of each individual component, rather than a complete bibliography inclusive of all relevant works to particular subtopics. However, important/milestone works for the geometric processing of images are included in this chapter.

### 3.1.3 Organization of the chapter

This chapter provides an overview of the geometric processing of images in the context of RBM. Three major components of geometric processing will be covered, including: (1) image geo-referencing, (2) dense image matching, and (3) texture mapping. Section 3.2 introduces a general framework/workflow that most of the RBM approaches are taking. Section 3.3, 3.4 and 3.5 will discuss the aforementioned three components in detail, and section 6 provides conclusions.

## 3.2 Reality-based modelling: A general overview

Reality-based 3D modelling (RBM) focuses specifically on representing the physical properties of objects, being the dimensions, type/class and topology (Verdie *et al.*, 2015). Initiated by the geo-community, its characteristics of being "real" drives many applications; to name but a few: wide-area urban wind and flood simulation; urban planning; solar and shadow analysis etc. (Biljecki *et al.*, 2015). Current practices in RBM in industry still rely heavily on manual intervention, and many entities have their own workflows, with variations mainly on measuring and identifying the topological structures and object elements.

Depending on the level of automation, the image-based RBM procedure can be broadly divided into automated and semi-automated/manual processing paths. Figure 3.1 illustrates common processing paths in both categories. Most RBM workflow starts with the geo-referencing of the raw data. The automated procedure (indicated by the red arrows) aims to automate the surface reconstruction, object identification/topological reconstruction with computer algorithms. The semi-automated/manual procedure (indicated by the green arrows) normally considers the 3D measurement, topological reconstruction and object/element identification as a process with the operator involved with every single object being modeled. The semi-automation comprises algorithms taking over part of the process including operations such as topological reconstruction (Gruen and Wang, 1998) from manually measured points or planes. With the reconstructed 3D models, the texture mapping procedure can be carried out by cropping the original images into textures and assigning each texture to the relevant face of the model (Hanusch, 2010).

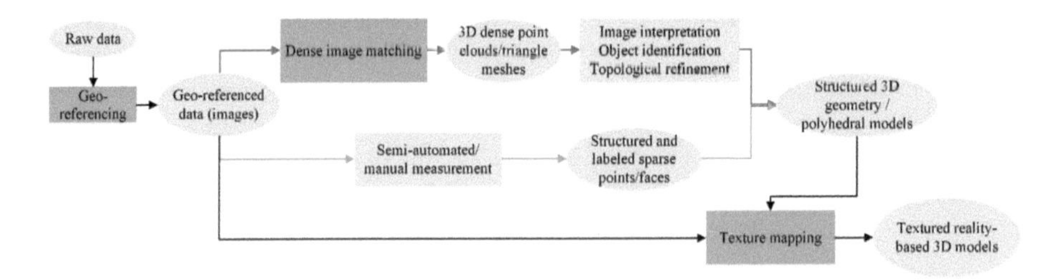


Figure 3.1 A general workflow of reality-based 3D modelling. Rectangles represent processing units and ellipses represent products (i.e. input, intermediate level results, final results). Light-blue rectangles represent the key components for geometric processing, and light green rectangles represent the object identification and topological reconstruction. The processing chain with red arrows indicates the automated workflow, and the path with green arrows indicates the semi-automated/manual workflow.

It should be noted that the taxonomy described using automation vs. semi-automation/fully manual is rather intuitive and does not encapsulate all circumstances, and sometimes the components of two processing paths (in Figure 3.1, outlined by red and green arrows) can be interchangeable. For instance, the measurement of key points for topological reconstruction can be aided by densely matched point clouds. In addition, automated methods are sometimes used in a semi-automated way: models generated by automated methods will go through a strict quality-control procedure, where manually identified errors will be corrected either through editing tools or ad-hoc correction algorithms (Xiong *et al.*, 2014). Improvement of RBM relies on the performance increase of each individual component (the rectangles shown in Figure 3.1) of the workflow. Each of them has its unique challenges, as follows:

*Geo-referencing:* The geo-referencing of images refers to the process of recovering the camera interior and exterior orientation parameters. It was usually a routine process in traditional photogrammetry for ordered images, where the manual intervention was usually in measuring tie points/GCP (Ground Control Point) or in gross error elimination. The major challenges are usually the lack of interest/tie points in homogenous regions, and the bundle adjustment with suboptimal camera network and GCP distributions. The state-of-the-art geo-referencing paradigm will be introduced in Section 3.3.

*Dense Image Matching*: Dense image matching (DIM) refers to the process of generating explicit 3D information (i.e. 3D point clouds, surface models). It requires a per-pixel level correspondence search across stereo or multi-stereo images for accurate 3D point determination. Although there has been much improvement for reconstructing reasonably dense 3D point clouds, the problem remains highly challenging for complex scenes with suboptimal camera networks. State-of-the-art DIM methods will be introduced in Section 3.4.

*Texture Mapping*: In the photogrammetry and computer vision domain, texture mapping refers to the process of mapping realistic textures from oriented images to the generated 3D geometry. If the 3D geometry is accurate, the process of texture mapping can be performed automatically. However, there are a number of challenges associated with it: (1) how to keep the textures seamless when textures are coming from different images; (2) how to deal with occluded areas; and (3) how to select the best set of images from multiple images for texture mapping. Details of current practice and methods for texture mapping are introduced in Section 3.5.

*Object identification and topological reconstruction*: Object identification and topological reconstruction is still largely a manual process in current industrial practice. Although there has been exciting progress in terms of image understanding (Krizhevsky *et al.*, 2012; LeCun *et al.*, 2015), current industrial practice still relies on a fully manual/semi-automated process. Information relating to object identification and topological reconstructions will be introduced in other chapters of this book as appropriate.

Advances in terms of geometric processing (the light-blue rectangular boxes in Figure 3.1) have evolved from a manually guided process (tie points measurement and blunder elimination in geo-referencing and manual terrain and surface measurement) to a largely automated/fully automated process; even non-experts are able to generate photorealistic meshes from images using state-of-the-art commercial and open source packages (Pix4D, 2017; Wu, 2014). Therefore in the following sections, we will focus particularly on current geometric processing methods from images, outlined by the three light-blue boxes in Figure 3.1.

## 3.3 Image geo-referencing

Image geo-referencing refers to the process of computing the interior and exterior orientations of the camera stations in a global or local coordinate system. This normally accompanies calibration of the interior camera parameters and lens distortions, being camera calibration. A parallel sub-field in computer vision (CV) that is highly relevant to geo-referencing is referred to as structure-from-motion (SFM), or pose estimation. Both CV and photogrammetry are very similar in terms of their mathematical foundations, while there are differences in terms of formulations: photogrammetry adopts algebraic operations at the scalar level for computations in Euclidean and central perspectives, while CV adopts homogeneous coordinates to represent the perspective geometry, and allows for flexibility in terms of affine distortion in the object space (Hartley and Zisserman, 2004). Photogrammetry, moreover, usually uses rigorous sensor modelling to achieve the highest accuracy for mapping/3D modelling, while CV has other aspirations in dealing with geometry, such as visual odometry/ego-motion, or robotics, where speed and robustness are also important such that approximate algorithms are often used. A more detailed comparison between these two fields is presented in Mundy (1993) and Remondino and Fraser (2006).

In this chapter, we follow photogrammetric convention, but occasionally include relevant information from computer vision for comparison. In general, a geo-referencing procedure refers to the steps in the unfilled rectangles shown in Figure 3.2. This is a general paradigm that most state-of-the-art geo-referencing approaches follow, with variations in methods used in individual or joints of the components.

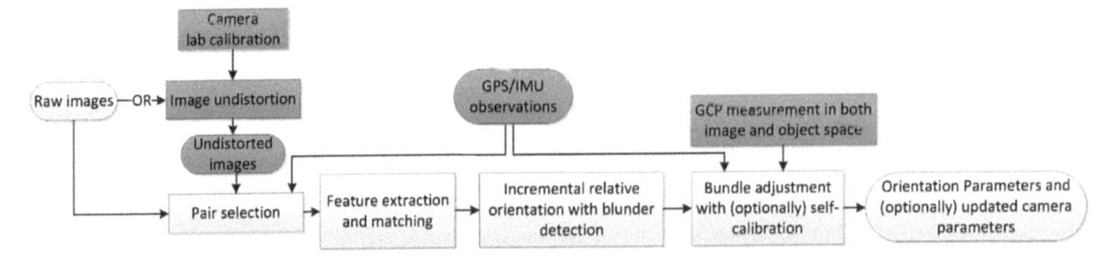


Figure 3.2 A general workflow for the geometric processing of images for 3D modelling. Unfilled rectangles represent processing units and ellipses represent inputs/outputs. The grey-fillled rectangles/ellipses are optional procedures that are case dependent.

The geometric processing workflow depicted in Figure 3.2 has existed for decades in the field of digital photogrammetry and has been consistently improved. Nowadays it has become the standard workflow for geo-referencing and pose estimation. The optional procedures (coloured grey in Figure 3.2) come with additionally available observations and laboratory calibration to improve the robustness of the workflow. Laboratory calibration of cameras was usually carried out in the calibration field with ground control array in traditional survey-based photogrammetry, while oftentimes it can be addressed as part of the bundle adjustment as self-calibration. GPS (global positioning system)/IMU (inertial measurement unit) observations can be add-ons for the image pair selection, and/or as prior information of the orientation parameters, and/or to provide the datum of the dataset. GPS/IMU observations for images are not always necessary but can be used to improve the efficiency and robustness of the workflow. GCPs can be introduced as additional constraints to provide the datum and correct potential ground topography deformation due to lens distortions. Both GCP and GPS/IMU, can be viewed as observations in a general bundle solution (Gruen and Beyer, 2001).

Given a set of images, the necessary geo-referencing procedure starts with *selecting pairs of images* and *extracting and matching sparse correspondences* to form an initial set of image observations, containing two or more ray points across multiple images. These image observations often contain many blunders (erroneous observations), and the *incremental relative orientation* aims to determine sparse corresponding points and eliminate blunders through statistical procedures as well as providing initial values for *bundle adjustment*. Bundle adjustment takes into account various observations, including image, GCP, GPS/IMU (approximations of exterior orientations), to accurately compute the interior and exterior parameters of images. We will provide a more detailed introduction for the components illustrated above.

### 3.3.1 Camera calibration

Camera calibration refers to the process of estimating camera interior orientations (focal length/ principal distance and principal points) and lens distortions. The goal of camera calibration is to separate the camera parameters such that it remains static when applied to other datasets (e.g. stereo systems), Calibration of metric cameras mainly refers to the determination of interior orientations, as lens distortions can usually be ignored. However, with the increasing use of consumer-grade cameras for metric calculation, the camera parameters (interior orientation and lens distortions) cannot be ignored. Nowadays, automated camera calibration with coded targets/natural image correspondences via bundle adjustment with self-calibration is becoming standard procedure, particularly for close-range photogrammetry, whereas the use of 3D calibration array is no longer mandatory in this case. Nevertheless, large camera systems, such as digital airborne photogrammetric systems, or integrated multi-sensor systems, still use 3D object array to estimate camera parameters, which again follows typical least squares adjustment through the well-known collinearity equations (Schenk, 2005). The calibration parameters normally comprise the perturbation terms in the image space as $\Delta x$ and $\Delta y$, being (Fraser, 2013; Gruen and Beyer, 2001):

$$\begin{aligned} \Delta x &= -\Delta x_p + \frac{\bar{x}}{c}\Delta c + \bar{x}r^2K_1 + \bar{x}r^4K_2 + \bar{x}r^6K_3 + (2\bar{x}^2 + r^2)P_1 + 2P_2\bar{x}\bar{y} + b_1\bar{x} + b_2\bar{y} \\ \Delta y &= -\Delta y_p + \frac{\bar{y}}{c}\Delta c + \bar{y}r^2K_1 + \bar{y}r^4K_2 + \bar{y}r^6K_3 + (2\bar{y}^2 + r^2)P_2 + 2P_1\bar{x}\bar{y} + b_2\bar{x} \end{aligned} \quad (1)$$

where $\bar{x} = (x - x_0)$ and $\bar{y} = (y - y_0)$ is the distance to the image center and $r$ is the radial distance, given by:

$$r^2 = \bar{x}^2 + \bar{y}^2 = (x - x_0)^2 + (y - y_0)^2 \quad (2)$$

$\Delta x$ and $\Delta y$ can be seen as the correction terms in the image space and as comprised of the correction of the principal points $\Delta x_p$ and $\Delta y_p$, the correction term of the principal distance $\Delta c$, the coefficients of radial distortion $K_i$, the coefficients of decentering distortion $P_i$, and the scale and affinity terms $b_1$ and $b_2$. Terms and parameter definition may vary slightly, while the major concept for modelling different types of distortion remains mostly static.

The ability of the camera to calibrate itself in bundle adjustment does not necessarily mean it works well in all situations. Critical aspects include the camera networks (sufficient orthogonal images), coverage of image observations across the image frame, depth variation in the scene, etc. It has been well understood that there are correlations between some of the interior and exterior parameters, e.g. principal distance with respect to perspective center, principal points with respect to decentering distortion, etc. This requires the adoption of highly convergent images and orthogonal camera roll angles to break some of these correlations (Fraser, 2013). Furthermore, the depth variation in the scene is also crucial to accurately recover the principal distance. These critical aspects may arguably improve the calibration approaches adopted by some computer vision systems, with the 2D chessboard being the primary calibration pattern (Zhang, 2000).

State-of-the-art photogrammetry software systems have adopted self-calibration as a standard technique to compensate for image distortions, sometimes caused by environmental conditions (e.g. humidity, pressure, etc.). Apparently, the images used for bundle adjustment may not be ideal for calibrating the system, which raises the probability of having to produce scene-dependent camera calibration parameters. Having said that, on-mission self-calibration with bundle adjustment may not be deemed simply as a black-box approach, and pre-calibration is necessary when the camera network, scene depth variation, and texture richness (for interest points extraction) are not optimal, and, in fact, is always suggested when it is possible (Remondino and Fraser, 2006).

### 3.3.2 Pair selection and tie point matching and extraction

Extraction and matching the tie points in the dataset is the most time-consuming part when dealing with a large image dataset. It normally starts with finding pair-wise correspondences. Selecting image pairs seems to be obvious in traditional aerial photogrammetry missions, as images are acquired following regular block structures, and most of the time the data come with the GPS/IMU observations. Nowadays, interest points can be extracted and matched fully automatically; only in extreme cases is it necessary to add tie points manually, such as areas with poor textures, or where existing tie points are unevenly distributed.

*Pair selection:* The use of GPS observations for pair selection in matching is rather straightforward, where neighboring images of an image defined by a pre-defined distance threshold, or *N* closest images given the GPS observations (Qin *et al.*, 2012). However, difficulties may arise when dealing with a large set of unordered images (close-range, crowd-sourcing images, and UAV (unmanned aerial vehicles) images with no GPS headers). Exhausting all possible pairs can be an order of magnitude slower (*N* times) than with ordered images. One of the strategies is to perform a preliminary exhaustive matching across all possible pairs at a much lower resolution to identify a set or cluster (Deseilligny and Clery, 2011). While it does improve computational efficiency, it may either omit important pairs, due to insufficient interest points, or return too many potential pairs due to scene similarity. Methods for clustering crowd-sourcing images were developed to group images that potentially cover the same scene content (Agarwal *et al.*, 2011; Frahm *et al.*, 2010; Havlena and Schindler, 2014; Li *et al.*, 2008): iconic scene graph (Li *et al.*, 2008), extract and cluster bag of words to build vocabulary trees (Nister and Stewenius, 2006) to cluster images, followed by geometric verification testing of the success of the fundamental matrix estimation. These graphs can be simplified to a skeleton and additionally use parallel computations to reduce the computation time (Agarwal *et al.*, 2011). Wu *et al.* (2013) reduced the computation at the level of pair-wise correspondence search, by only matching the first 100–400 points ranked by their scale of SIFT (scale-invariant feature transform). Lowe (2004) features initializing the connectivity graph, which could dramatically reduce the pair-wise matching time. It may be worth noting that these strategies for pair selection utilize the information content of the images, and this is at the cost of discarding pairs that do not pass initial tests, as well as reducing the pairs being used for reconstruction (skeleton of connectivity graph), with the potential risk of generating highly singular camera networks (poor convergence and overlap). Therefore, with close-range images used for generating high-precision models, strategies for reducing the computation while maintaining good camera networks will be carefully revisited.

*Feature extraction & matching*: Feature correspondences across multiple images provide ties as the observations for recovering the geometric parameters of the images. Features mostly refers to image points in practice, although there are works incorporating lines as the observations (Habib *et al.*, 2002). Extraction and matching are two distinctive steps: (1) a set of interest points is extracted in each individual image and (2) the matches of the points are then carried out to generate correspondences. There are mainly two types of points extracted by point operators: (1) corner points and (2) blob points. Methods designed for detecting corner points usually look for points with large gradients in at least two intersected directions, as opposed to one direction (refers to lines). Examples of corner-based detectors are Moravec (Moravec, 1980), Harris (Harris and Stephens, 1988), Förstner (Förstner and Gülch, 1987), as well as their variations. An area-based point detector does not necessarily locate the point at corners: rather, it considers local extrema of an area in the scale space to achieve the scale-invariance. Detectors in this category include operators that use Difference of Gaussians (DoG) or Laplacian of Gaussian (LoG) to define the interest points, e.g. SIFT (Lowe, 2004) and its variations such as affine-SIFT and PCA-SIFT (Ke and Sukthankar, 2004; Morel and Yu, 2009), and similar approaches such as SURF (speed-up robust features) (Bay *et al.*, 2006). There are also detectors that detect both types of points (Förstner *et al.*, 2009). The detectors are generally designed to find scale- and rotation-invariant features, such that they appear repetitively with image contents being scaled and rotated. These detectors extract distinctive points or points with rich surrounding textures, such that these points across different images can be matched.

Feature matching often starts with two images, where every point on one image is compared with each of the points on the other image; this apparently takes $O(n_1 n_2)$ complexity, where $n_1$ and $n_2$ are the number of feature points in each of the images. Many matching methods have been proposed in the past: one class of algorithm compares the surrounding textures of each point, such as sum of squared differences (SSD) (Kanade and Okutomi, 1994), normalized cross correlation (NCC) (Lewis, 1995), and least squares matching (LSM) (Gruen, 1985). Another set of algorithms extracts feature vectors associated with the points as feature descriptors, and comparisons are made directly between feature descriptors. One of the most famous descriptors (and a milestone work) was the SIFT descriptor (Lowe, 2004), where the histogram of the gradient vector at the scale space was compacted as a feature descriptor, which has advanced matching correctness to a notable level, being used in many state-of-the-art systems, e.g. Bundler (Snavely, 2010), visualSFM (Wu, 2014), Apero (Deseilligny and Clery, 2011), and triggering subsequent works utilizing similar ideas.

However, there are still many unsolved problems: lack of speed performance for feature extraction and matching is still one of the major challenges in the geometric processing pipeline. Therefore, consolidated efforts at increasing speed performance without a corresponding loss of matching performance is worth investigating. Furthermore, sometimes such operators generate too many points/matches on the texture-rich area (sometimes only a few pixels apart) leading to unnecessary computations; hence strategies to reduce the redundant points are also important for generating well-suited observations for geometric processing. Lastly, very often the localization of area-based point detectors is not accurate: as shown in Remondino (2006), SIFT point detectors have accuracies of only 2–3 pixels, which potentially creates extra uncertainties in the observation. A recommended procedure in high-precision georeferencing is to use corner detectors (e.g. Harris) and SIFT descriptors to perform an initial correspondence matching, and then perform a refinement matching using least squares image matching to achieve high accuracy; however, this will be at the expense of increasing the computation complexity.

### 3.3.3 Incremental relative orientation with blunder detection

During a decade of development, incremental relative orientation (RO) is now becoming the standard strategy for 3D reconstruction, along with its derived strategies (Frahm *et al.*, 2010; Gherardi *et al.*, 2010; Wu, 2013). The key role of incremental relative orientation is to eliminate blunders of tie points and provide initial values of the exterior orientation parameters for bundle adjustment. This is a fully automatic procedure that integrates feature correspondences across multiple images and robust statistical strategies (RANSAC – random sampling consensus). Incremental relative orientation refers to the process starting with a two-view relative orientation, followed by sequentially orienting the rest of the images given the feature correspondences. Relative orientation was developed a long time ago, and the only barrier to automation is the blunder elimination (gross errors) for wrongly matched points. Until the use of RANSAC, the major technique was to eliminate points with large residuals in relative orientation (residual-based blunder elimination); this is very sensitive to the noise ratio of the feature matching. RANSAC uses a random sampling strategy that starts with randomly sampled feature matches (observations) instead of all the observations for relative orientation (model estimation), runs the same process multiple times and selects the model (estimated orientation parameters) accounting for most of the observations with reasonable residuals. This has dramatically improved the automation in relative orientation and subsequently the incremental procedure, as theoretically it only requires that the error rate of the matches be larger than 50%; apparently, state-of-the-art feature extractors and matchers do much better with images in most applications.

The RO itself in photogrammetry is comparatively static, where orienting two images relatively (with arbitrary datum) requires a minimal five corresponding points, normally formulated as a least squares problem solved with Newton methods (Mikhail *et al.*, 2001; Schenk, 2005). In computer vision the computation is carried out through an estimation of the fundamental/essential matrix (Hartley and Zisserman, 2004) and the DOF (degree of freedom), which vary with the constraints on the matrix, being 8-point, 5-point, 4-point and 3-point algorithms (Fraundorfer *et al.*, 2010) (with known angles or translation). The RO problem in CV is normally solved using SVD (singular value decomposition), with additional constraints normally formed in a mathematical sense. Essentially, the flexibility of estimating the fundamental matrix instead of a rigorous formulation (with unknowns represented by angles and translations) allows geometric reconstruction in an affine space. When it comes to the relative orientation for metric purposes, the rigorous formulation with iterative solutions normally delivers more accurate results and less correlated parameters.

### 3.3.4 Bundle adjustment

Bundle adjustment refers to the process of computing the unknowns from the collinearity equations that contain all the observations. This step is taken following incremental relative orientation, aiming at correcting the post-estimation drift of the incremental orientation and re-adjusting the unknowns under a least squares framework. Nowadays, this is rather standard in photogrammetry and computer vision. In computer vision it generally refers to the minimization of the re-projection errors, being measured image points from feature correspondences (observations) as compared to the same points predicted by the bundle system (Hartley and Zisserman, 2004). In photogrammetry, the observations refer to a wider spectrum of prior information in addition to the feature correspondences, including ground control points (GCP), some of the exterior orientation parameters, or GPS/IMU observations. These observations are associated with a weight matrix accounting for the individual contributions of these observations in the bundle system. Following the formulation of Gruen and Beyer (2001), the observational equations can be represented as follows:

$$\begin{aligned} -e_B &= A_1 x_p + A_2 t + A_3 z - l_B \quad ; \quad P_B \\ -e_p &= I x_p \qquad\qquad\qquad\; - l_p \quad ; \quad P_p \\ -e_t &= \qquad\quad It \qquad\qquad - l_t \quad ; \quad P_t \\ -e_z &= \qquad\qquad\qquad Iz - l_z \quad ; \quad P_z \end{aligned} \tag{3}$$

where $A_1$, $A_2$ and $A_3$ are the design matrix (Jacobian matrix) associated with the object point $\boldsymbol{x}_p$, orientation parameters $\boldsymbol{t}$, and the additional camera parameters (calibration parameters) $z$. In Gruen and Beyer (2001) this refers to the 11-parameter model. $e_b$, $e_p$, $e_t$ and $e_z$ are the error terms, and $\boldsymbol{l}_B$, $\boldsymbol{l}_p$, $\boldsymbol{l}_t$ and $\boldsymbol{l}_z$ are the observation vectors for image observations, object space point observations (GCP), orientation parameter observations and prior knowledge about the camera calibration parameters (such as from lab-based camera calibration), respectively. $P_B$, $P_p$, $P_t$, $P_t$ are a priori weights measuring the confidence of the measurement, usually correlating to the standard deviation of the observations. The weights for image observations (tie points) are normally static and this depends on the localization accuracy of the feature extractor, while it might be adjusted for tie points measured manually (for images with weak connections). The second to the fourth equations of (3), correspond to the observations directly related to the unknowns. For GCP (the second equation of (3)), the weights are usually set large to enforce hard constraints on the accuracy of GCPs, and they are often adjustable due to the fact that GCPs coming from different resources (directly measured GCP, or derived GCP from other sources) have different accuracies, and these can be well reflected in the a priori weight. A similar idea applies to the observations of the orientation parameters: they can either be oriented parameters for some of the images (Qin, 2014), or observations from GPS/IMU data. The observations for additional parameters (camera parameters) are normally used when the camera has been precisely calibrated. Based on a similar concept, extensions of the bundle adjustment with multiple camera systems should be straightforward.

The design matrix ($A_1$, $A_2$ and $A_3$, practically will be merged as one) is highly sparse. Past research work has been carried out in optimally ordering the elements that enable efficient solvers utilizing the sparsity of the matrix; many cases were discussed, including block-wise ordering, banded matrices (Brown, 1976), and formulations that cancel the object-space unknowns to gain efficiency (Deseilligny and Clery, 2011). Efforts in paralleling the bundle adjustment for multi-core CPU and GPU computation are also worth noting (Wu *et al.*, 2011), providing potential paths to real-time bundle adjustment. Additionally, efforts made initially as for relative orientation and bundle adjustment – sequential bundle adjustment (Grün, 1985) are nowadays widely used in the computer domain in visual odometry (Nistér *et al.*, 2004). The idea of sequential adjustment is to perform bundle adjustment in the process of incremental relative orientation; instead of starting everything over, it directly edits the element of the design/normal matrix to update the necessary elements to save computational time. Therefore, although bundle adjustment varies with different applications and demands, the mathematical basis remains rather static. For well-planned data acquisition for 3D modelling, the process is largely a routine process.

## 3.4 Dense image matching

Dense image matching (DIM) refers to the process of searching for dense pixel correspondences across oriented images. Being 'dense' is relative to the sparse tie points used for image orientation, often order of magnitude higher than the density of tie points (or interest points). Although not definitive, DIM nowadays normally refers to per-pixel correspondence search (Scharstein and Szeliski, 2002). Given the image orientation, epipolar constraints are used to reduce the 2D correspondence search problem to 1D. Despite this simplification, the DIM problem still remains a challenge that attracts a great deal of attention from the computer-vision and geo-community (Scharstein and Szeliski, 2014). Depending on the number of images used in matching, DIM can be classified simply as stereo image matching or multi-stereo/multi-view image matching.

### 3.4.1 Stereo image matching

A stereo pair (or binocular image) refers to two images taken from different perspectives. The task is to find for each pixel in one image, its correspondence in the other. With the epipolar constraint, a popular convention is to rectify the pairs to the epipolar image space such that correspondences are in the same row between two images, and thereby the matches can be represented by the column coordinate differences, termed disparity or parallax. For a match with $(y_l, x_l)$ and $(y_r, x_r)$, being the left and right pixel coordinate with a disparity $d$, equation (4) follows:

$$y_r = y_l,\ x_r = x_l - d \tag{4}$$

Stereo image matching algorithms normally operate on the epipolar images with the goal of generating pixel-wise disparity images (corresponding to either left or right view). Despite a multitude of solutions for this well-investigated problem, the

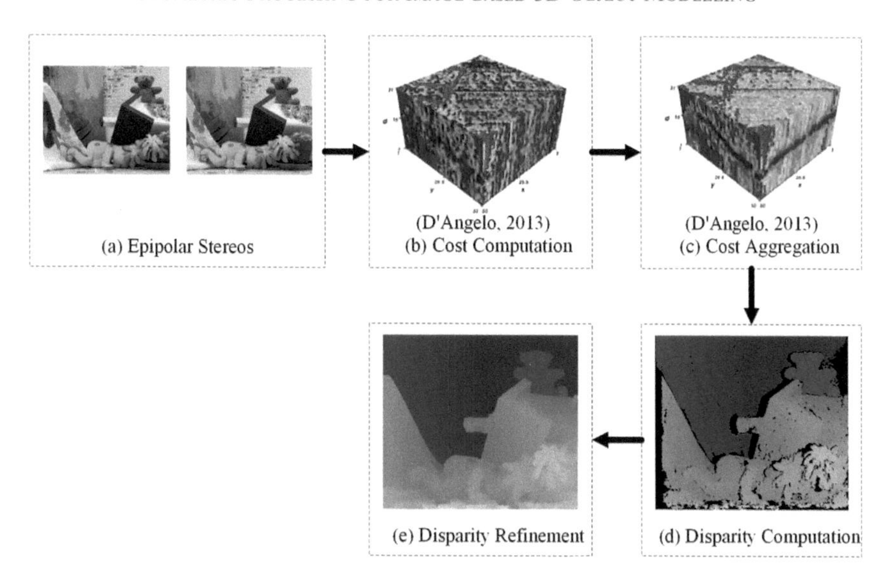


Figure 3.3 The workflow of rectified stereo image matching. Details are described in the text in a sequential order.

algorithms can normally be generalized in a 4-step workflow (Scharstein and Szeliski, 2002): (b) cost computation (c) cost aggregation (d) disparity computation and (e) disparity refinement, as shown in Figure 3.3.

Stereo matching can largely be framed as a minimization problem $\underset{D}{\operatorname{argmin}} E(D, I_1, I_2)$, which tends to find the optimal disparity map $D$ of the epipolar images $I_1$, $I_2$ by minimizing an energy function $E$. The energy is often formulated through the measure of coherence/similarity between corresponding pixels indexed by $D$, e.g., per-pixel squared colour differences aim to minimize the sum of the squared error. Very often a simple per-pixel colour/feature difference, also referred to as cost, is not sufficient to yield accurate disparity maps (Duan *et al.*, 2016). Therefore, a priori assumptions are normally posed to the disparity image itself, such that the disparity, or the 3D surface, is smooth and does not vary much between neighboring pixels. This leads to regularizations of the $D$ in the energy function such that the neighboring pixels are intercorrelated. This is normally processed through a step called cost aggregation, whereby the initial per-pixel costs computed independently are inferred and aggregated following a certain strategy to realize the regularization. The cost aggregation serves as the key to produce spatially coherent surfaces and a disparity map. An optional refinement may also be carried out to filter out or incorporate additional information to further enhance the disparity map.

#### 3.4.1.1. Cost computation

The cost defines the possibility of two pixels being a correct match through measuring their similarities, also called the cost metric. A small 'cost' value indicates a good chance of match and vice versa. In stereo matching, the 'cost' is computed for every potential match (across a disparity range), thus generating a cost volume with a dimension $W \times H \times L$ (Figure 3.3b) $C \sim \mathbb{R}^2 \times \mathbb{R}$. Examples of the cost metrics can be either pixel-based metrics such as absolute difference (AD), gradient differences, the insensitive measure of Birchfield and Tomasi (BT) (Birchfield and Tomasi, 1998), or window-based metrics such as zero-based normalized cross correlation (ZNCC), normalized gradient (Zhou and Boulanger, 2012), Census (Jiao *et al.*, 2014; Kordelas *et al.*, 2015; Zabih and Woodfill, 2005) and mutual information (MI) (Paul *et al.*, 1997). Given the large variation in radiometric properties for stereo pairs, these metrics may perform significantly differently. Usually, pixel-based methods are sensitive to noise and radiometric discrepancies. AD, BT or the gradient measures usually work well on images with very similar radiometry, while they generally generate errors at the presence of additive or multiplitive radiometric differences. Window-based methods are normally used to account for such radiometric differences: e.g. ZNCC or NCC computes the correlation between pixel windows, which is invariant to additive radiometric differences. More advanced window-based methods, such as Census and MI, are able to account for non-linear radiometric differences: the Census metric adopts non-parametric transformations that utilize the rank of the pixels within a window, while MI computes the statistical correlations between pixel windows. Such formulations naturally account for non-linear variations of pixel values. Hirschmüller and Scharstein (2009) performed a systematic study evaluating these cost metrics and concluded that Census and MI measures can achieve the best matching results under varying testing conditions. However, the drawback of these methods is that they are highly sensitive to image noise, particularly the Census metric.

More recent cost metrics have adopted machine learning approaches, where large numbers of sample pairs of stereo images with ground truth disparity are learned through a convolutional neural network (CNN) (Žbontar and LeCun, 2015, 2016) and applied to other images pairs to produce the confidence of matches. A pre-eminent fact is that, among the top performed algorithms in the open source public dataset, most of them used cost metrics generated through CNN. However, a potential issue for this method is high demand for training data and that it requires large computational resources. Moreover, its transferability to real and large-scale reconstruction still needs to be validated.

#### 3.4.1.2 Cost aggregation

Taking matches with the smallest cost metrics normally introduces significant errors, largely due to texture-less areas in the image and texture repetitiveness. The idea of cost aggregation is to pose smoothness constraints by assuming that scenes are mostly piecewise smooth (Scharstein and Szeliski, 2002). Such smoothness constraints can be posed either on the disparity or the normal of the planes. Algorithms using disparity smoothness constraints are called *1D label algorithms*, since for each pixel the algorithm assigns one label. Normal vector smoothness constraints usually operate on meshes or detected planes, by assuming adjacent meshes with similar colour in order to keep the similar normal vectors. It is also called *3D label algorithms*, since three labels (disparity and normal direction) are assigned for every pixel.

*1D label matching*: In general, 1D label matching can be classified as local matching and global/semi-global matching. Local matching assumes that all pixels with

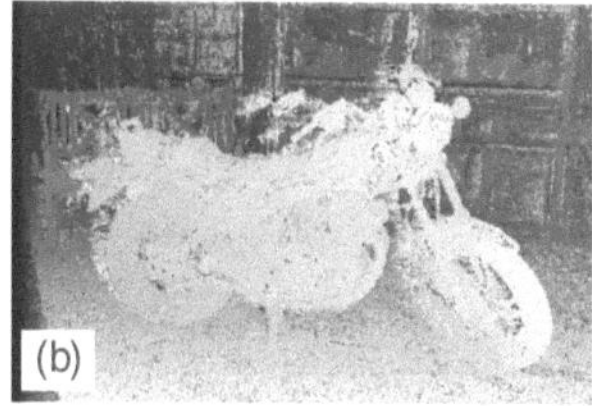

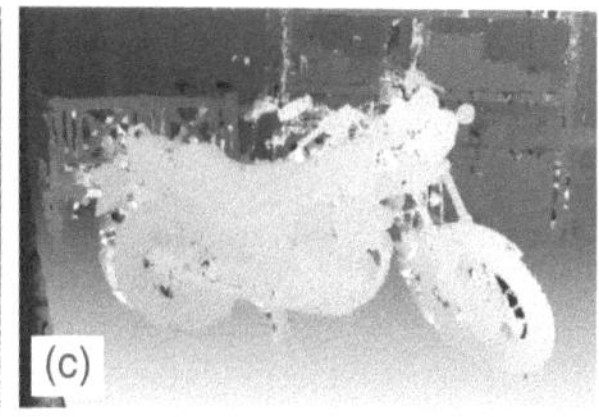


Figure 3.4 Disparity images before/after bilateral filter: (a) original image; (b) disparity image after cost computation; (c) disparity image after bilateral filter.

similar colour in the support window should have the same disparity. They aggregate cost in a local small window. Examples of local matching algorithms include bilateral filter (Yoon and Kweon, 2006), image-guided filter (He *et al.*, 2013) and minimum spanning tree filter (Yang, 2015). These methods are weighted filters through the cost volume (Rhemann *et al.*, 2011). Local matching is simple and fast, and it normally reduces salt-and-pepper errors and the state-of-the-art filters are edge-aware; however, it goes easily into local minima of the energy function and is not generally robust in the entire region (Figure 3.4 shows speckled noises in the results).

Global/semi-global matching methods tend to aggregate all the pixels instead of the local support window. This essentially imposes a regularization term (also called a smooth term) over all the pixels in the energy minimization problem:

$$\underset{D}{\operatorname{argmin}}\, E(\boldsymbol{D}) = E_{data}(\boldsymbol{D}) + E_{smooth}(\boldsymbol{D}) \tag{5}$$

where $E_{data}$ is the cost term computed from the stereo data using cost metrics, and $E_{smooth}$ regularizes the entire disparity (rather than a support window) being smooth, wherein $E_{smooth}$ can effectively be a term that minimizes the first- or even higher-order derivatives of $\boldsymbol{D}$. However, this problem is an NP-hard one that often requires approximate solutions. Popular solutions include graph cuts (Kolmogorov and Zabih, 2001; Papadakis and Caselles, 2010; Taniai *et al.*, 2014), belief propagation (BP) (Yang *et al.*, 2009), dynamic programing (DP) (Hirschmüller, 2008). Graph cuts formulate this problem as a graph-labeling problem, while belief propagation solves this problem under a general Markov random field model. Both are iterative algorithms that tend to infer disparity consistency across the entire image; therefore they are regarded as global methods. Dynamic programming operates the inferences through lines across the image grid (rather than the entire image grid), and the solution itself is not iterative; therefore it is regarded as a semi-global method.

Among existing solutions, a variant of the DP method, the semi-global method (SGM) (Hirschmüller, 2008), has proven to be one of the best performing algorithms in terms of its accuracy and efficiency. The idea behind this method is that of sequentially inferring disparities through multiple directional lines using DP (Figure 3.5a). The advantage of utilizing multiple directional lines (or paths) is that it overcomes disadvantages of the single line DP/BP methods, which create serious streak errors (compare images b and c in Figure 3.5), and less global smoothness.

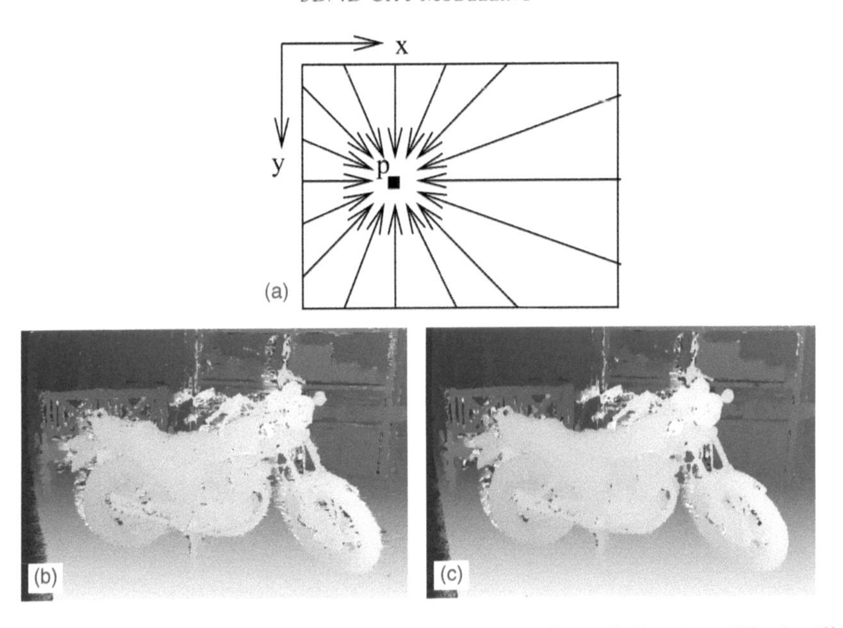


Figure 3.5 Cost aggregation principle of SGM: (a) 16 paths from all directions (Hirschmüller, 2008), where the DP method is applied in each direction and summed; (b) cost aggregation with only one path; (c) cost aggregation from all directions.

Compared to global approaches, SGM is extremely efficient and can be applied to large format images. It can achieve the goal of quasi real-time matching with the help of GPU. The complexity of SGM is $O$ ($WHL$) ($W$: Image width, $H$: Image height, $L$: Disparity range). Although no longer the best matching method in the Middlebury test (Scharstein and Szeliski, 2014), given its trade-off in efficiency and accuracy, SGM strategy is still the best performing approach and has been widely implemented by commercial software packages such as SURE and PhotoScan, ERDAS, Smart3D, etc.

1D label methods assume fronto-parallel planes, meaning the algorithms work best for planes that are parallel to the baseline. Ignorance of slant planes leads to the known 'staircase' effects; examples are shown in Figure 3.6 (c–g), computed by state-of-the-art 1D labeling methods. 3D labeling methods consider the normal of the pixels in a 3D context, by regularizing the normal direction of neighboring pixels in addition to the disparity value (three labels, as the length of the normal is not considered) (Olsson *et al.*, 2013). Algorithms have demonstrated that the idea of introducing the normals can effectively address this 'staircase' problem, e.g. the PMSC algorithm (Li *et al.*, 2016)) (Figure 6h). Given the increased per-pixel unknowns, solving the energy function is an ill-conditioned problem, and thus almost all 3D label algorithms assume that the scene is piecewise continuous, where normals are only computed for each segment (presuming the segmentation is performed prior to the energy minimization), greatly reducing the number of unknowns. Considering the energy minimization problem in equation (5) $E_{data}(\boldsymbol{D})$, refers to the initial

3D label matching

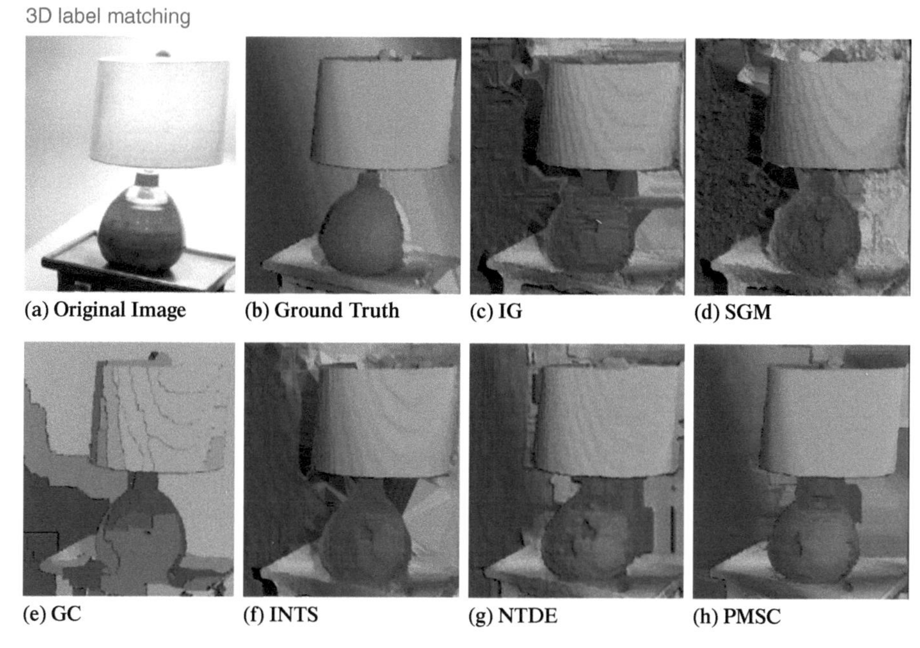


Figure 3.6 Results of different matching methods in slanted planes (Huang *et al.*, 2017): (c) Image-guided matching (Pham and Jeon, 2013); (d) Semi-global matching (Hirschmüller, 2008); (e) Graph cuts (Kolmogorov and Zabih, 2001); (f) Image-guided non-local matching with three steps (Huang *et al.*, 2016); (g) Non-textured region and the denoised edge map based matching (Kim and Kim, 2016); (h) PatchMatch-based superpixel cut (PMSC) for stereo matching (Li *et al.*, 2016).

cost of matches that can be constructed similarly as 1D labeling problem (irrelevant to the normals), with the basic unit being patches/segments, e.g. the cost can be computed as the average similarity measures of pixels in the patches. $E_{smooth}(\boldsymbol{D})$, however, is different, as the normal will be taken into account: in addition to imposing smooth constraints on the disparity value, it imposes smoothness constraints on the normal direction of patches. There are usually two constraints for adjacent patches: *connectivity* and *coplanarity*. Connectivity panelizes disjoint surface patches in the 3D space (like disparity jump, but for the patch) (Figure 3.7a–b), while coplanarity penalizes adjacent patches with sharp angles (non-coplanar) (Figure 3.7c–d).

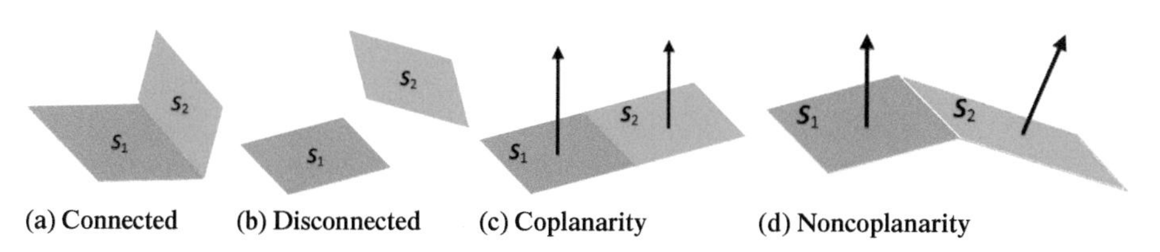


Figure 3.7 Patch-wise constraints for 3D label regularization.

The formulation of such constraint is straightforward (Taniai *et al.*, 2014; Zhang *et al.*, 2015), and some of the approximating methods used in 1D labeling problems can be readily applied, e.g. graph cuts (Bleyer and Gelautz, 2005), belief propagation (Guney and Geiger, 2015; Klaus *et al.*, 2006; Yamaguchi *et al.*, 2012), minimum spanning tree (Veldandi *et al.*, 2014), least squares (Huang *et al.*, 2017) and scan line algorithm (Barnes *et al.*, 2009; Li *et al.*, 2016; Zhang *et al.*, 2015). The reconstructed surface of 3D label matching methods are normally higher in accuracy for piecewise surfaces. Given the higher complexity, the computational time is also longer than similar algorithms in 1D labeling. Furthermore, since the methods use segmentation to generate patches, errors from segmentation (e.g. under-segmentation) may introduce additional mismatches in the final disparity.

#### 3.4.1.3. Sub-pixel interpolation and consistency check

Sub-pixel interpolation is becoming a standard processing for disparity calculations. Normally the labeling space contains only a finite number of integers (measured by pixels). The sub-pixel interpolation tries to fit a quadratic function using disparity and its corresponding costs (or aggregated cost) to find the decimal disparity value that corresponds to the valley of the quadratic function. Another common process is to use a left-right consistency check to eliminate mismatches: it calculates two disparity maps by switching left and right images, and compares their consistencies. Inconsistency pixels will be eliminated, and these pixels are largely occluded pixels.

#### 3.4.1.4. Disparity refinement

Many algorithms include a post-process directly on the disparity maps using image processing techniques. Such a process is often advantageous in generating clean disparity/3D surfaces. Common post-processing techniques include: (1) Speckle filter; (2) Weight median filter; (3) Intensity consistent disparity selection; and (4) Patch-based refinement. *Speckle filter* (Hirschmüller, 2008) removes an isolated disparity region if it is smaller than a given threshold. *Weight median Filter* is similar to the classic median filter, the only difference being that it uses the intensity values to infer the ranking of the pixels to obtain edge-aware results (Mozerov and Weijer, 2015). *Intensity consistent disparity selection* is a post-segmentation method that fits the best plane for each segment of the image (Hirschmüller, 2008), which is effective for correcting matching errors in the textureless region. The *patch-based* method (Drouyer *et al.*, 2017) follows a similar strategy by defining an optimal plane for the patches; the difference is that it also allows the implementation of smooth priors to adjust neighboring planes as the 3D labeling algorithm.

### 3.4.2 Multi-stereo/multi-view image matching

Typical mapping tasks contain much more than two overlapping images, e.g. aerial survey, unmanned aerial vehicles (UAV) data and close-range images. Multi-stereo/multi-view image matching methods are used to reconstruct 3D surfaces from a full set of overlapped images. With additional data, such as LiDAR or GIS data, matching algorithms can be customized to achieve better performances (Andreasson *et al.*, 2011; Diebel and Sebastian, 2006; Wang and Ferrie, 2015; Yang *et al.*, 2007).

#### 3.4.2.1. Multi-stereo matching

Multi-stereo matching is a direct extension of stereo matching by dividing multiple images into different pairs. Surfaces/depth maps are then generated for each pair using stereo matching algorithms (section 3.4.1), followed by a fusion step to generate the final 3D surface. In addition to the stereo matching algorithm, the critical issues of this class of method are (1) image pair selection, and (2) pair-wise fusion.

*Optimal image pair selection*: Pair-selection on images with regular photogrammetric acquisition is normally easy to perform, as neighboring images can be selected given the perspective center position as if in a 2D plane (Graça *et al.*, 2014; Yuan, 2008), and various parameters such as convergence and overlaps have been taken into consideration. However, for unordered images, such as mobile images, or close-range, and even some irregularly flied UAV images. Optimal pair selection is important, as over-selection may significantly increase the computational time, while under-selection may results in incomplete surfaces. Assuming each image as a node, a prevalent strategy is to formulate the pair selection as a graph analysis problem, where parameters of interest – such as intersection angle, baseline, resolution and repetition (Tao, 2016) – are set as the optimizing goal for the graph to make pair connections (Furukawa *et al.*, 2010).

*Pair-wise fusion*: The fusion of the pair-wise results helps complete the final 3D surface, as well as eliminate potential blunders and improve the accuracy using redundant observations. Based on the intermediate product of stereo matching, the fusion can act in meshes, point clouds or depths. *Mesh based fusion* aims to break and merge the topology of the meshes to form a complete and non-repetitive surface mesh (Newcombe *et al.*, 2011; Turk and Levoy, 1994). This includes complex mesh operations that perform triangle clipping, reconnection, as well as the consideration of eliminating spiking noises. *Volumetric fusion* methods vote and regularize point clouds, or surface meshes in 3D volumetric spaces. Statistics of the points/meshes in each volumetric unit are to determine the optimal presence of the 3D surface (Koch *et al.*, 1998; Sato *et al.*, 2002). Such a method is normally easy to implement; the downside of it may be the potential memory demand and aliasing effects created by regular space sampling. *Depth fusion* is one of the most flexible types of fusion method, due to its ease of implementation. It requires a common plane that aligns the candidate depth maps, thus various image based fusion algorithms (such as bilateral filter) can be used. This method is best practiced in mapping related fusion, as the ground plane can be simply regarded the common plane for fusion (Wenzel *et al.*, 2013).

#### 3.4.2.2. Multi-view matching

Multi-view matching uses multiple images simultaneously in matching. This class of algorithm considers the evaluation of multi-ray intersection (more than two) either locally or globally. The increased redundancy normally yields more robust and accurate results. With multiple images, it is generally not possible to operate the matching process in the rectified stereo space. Instead, the matching process is performed in the 3D space, either under the representation of 3D volumetric space, or the vertical line space. In general, multi-view matching methods can be categorized into three classes: vertical line locus (VLL) based matching; volumetric matching; and rough surface evolution matching.

*VLL-based matching*: The searching space of one pixel is defined in its ray direction, parameterized with either elevation or depth, and the matching score is computed through similarity measurement on re-projected pixel position on other images, as shown in Figure 3.8. A potentially selected match can directly return the 3D coordinate (Duan *et al.*, 2016; Furukawa and Ponce, 2007; Zhang, 2005). Given that the images are not rectified in the epipolar space, the similarity measurement will need to account for potential rotations and distortions of corresponding pixel windows. However, due to occlusion, noises, weak and repetitive texture, specular reflection, etc., the independent per-pixel measurement will likely induce errors. Thus, smoothness constraints between adjacent pixels should be considered. Similar to stereo matching, a global energy function with regularization terms can also be constructed, and the solution to minimizing the energy function can be obtained through graph cuts (Kolmogorov and Zabih, 2002), VLL-based SGM (Zhang *et al.*, 2017), etc. VLL

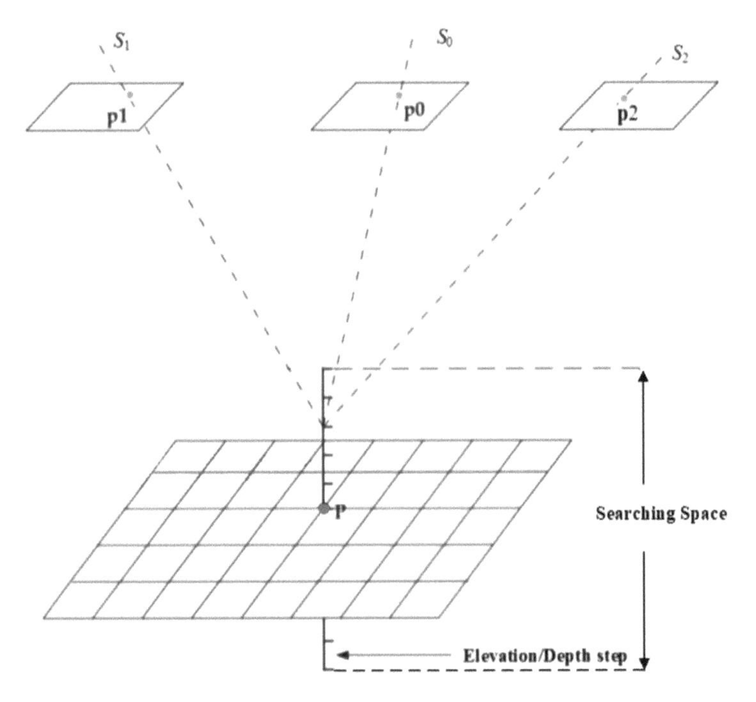


Figure 3.8 Principle of VLL-based matching (Zhang *et al.*, 2017). $S_0$, $S_1$ and $S_2$ are multi-view images where point P is potentially visible. p0, p1 and p2 are re-projected pixels on $S_0$, $S_1$ and $S_2$, respectively. The searching space is often pre-defined from artificialities, initial feature matching or initial DEM/DSM products.

methods are image-space methods, meaning that they generate point clouds for every image; thus a point clouds fusion step will be considered for multi-view matching to generate complete 3D data.

*Volumetric matching*: this method defines a regular 3D grid to represent the entire objects. The basic units of the grid, being voxels, are assigned binary labels {"Object", "Background"} to represent the presence of the object. The goal of volumetric matching is to find an optimal partitioning of the 3D grid into two segments (object and background). The boundary between the two regions is the object surface. This approach includes space carving (Broadhurst *et al.*, 2001; Kutulakos and Seitz, 2002), level-set (Faugeras and Keriven, 1998; Pons *et al.*, 2007) and volumetric graph cuts (Lempitsky *et al.*, 2006; Vogiatzis *et al.*, 2007). Space carving recovers the surface of objects with maximal photo-consistency. However, mismatches often occur in weak texture regions. Both level-set and volumetric graph cuts aim to recover regularized surface with optimal photo consistency (Boykov and Lempitsky, 2006). Volumetric matching methods can obtain complete object surfaces in true 3D by integrating multi-view images, and do not require point cloud fusion. However, such methods are normally limited by the large memory demand and the aliasing problems, and solutions for such problems are attempted by multi-resolution methods (Blaha *et al.*, 2016).

*Surface evolution matching*: This class of method refines vertices or meshes on initially generated rough object surfaces to more accurate, smooth and complete object surfaces iteratively (surface evolution). Initial surfaces can be obtained by quasi dense matching, e.g. PMVS (Furukawa and Ponce, 2007); plane sweeping (Vu *et al.*, 2012); geometrically constrained cross-correlation ($GC^3$) (Zhang, 2005); triangle matching (Zhu *et al.*, 2010), space carving (Broadhurst *et al.*, 2001; Kutulakos and Seitz, 2002); and silhouette reconstruction (Matusik *et al.*, 2000). Initial surfaces provide approximate surfaces of the objects, and the optimal surfaces can be obtained by minimizing an energy function with multi-view photo consistency constraints and variational surface smoothness constraints (Slabaugh and Unal, 2005; Tyleček and Šára, 2010; Vu *et al.*, 2012). The energy function is defined in a continuous infinite domain. Given good initial values, the gradient of the energy function points to the optimal solution, and the refined surfaces are getting closer to the optimal surfaces within a reasonable number of iterations. With the help of a graphics card, it is possible to reconstruct large-scale outdoor scenes with high accuray and reasonable running time (Vu *et al.*, 2012).

### 3.4.3 Joint geometric and semantic estimation

A recent trend tends to combine retrieving object labels and recovering geometry using multiple images into a single optimization framework (Häne *et al.*, 2013; Savinov *et al.*, 2016). The underlying concept of this class of method is that both tasks can mutually enhance one another. This essentially becomes an extension of the volumetric method, as the object label is regarded as one of the properties of each volumetric unit. This combination has proven to be successful in achieving good geometric and semantic results (Guney and Geiger, 2015; Ladický *et al.*, 2012; Yamaguchi *et al.*, 2014). Again, this formulation can be constructed as an energy minimization problem

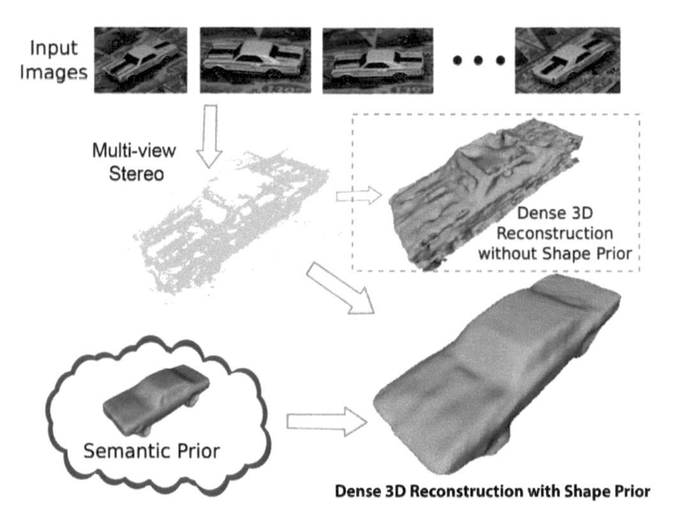


Figure 3.9 Semantic prior constrained reconstruction (Bao *et al.*, 2013).

(Savinov *et al.*, 2016) and the semantic labels can be initialized using a simple image-based classifier. In addition to estimating the object types, priors of the object shape can also be incorporated (Bao *et al.*, 2013; Wei *et al.*, 2014): Given training data comprised of accurate 3D models and multiple images, a semantic prior can be learned to describe the general shape of the category to constrain the multi-view matching (Figure 3.9).

## 3.5 Texture mapping

Texture mapping refers to the process of assigning patches of original images to the polyhedral models and triangle meshes. It can increase the visual appearance and realism of 3D models and provide additional information for visual interpretation (e.g. patch classification and semantic labeling). In RBM, with 3D geometric models generated from oriented images, it appears to be straightforward to assign and crop the same set of oriented images to each polygonal/triangle face (we use 'face' thereafter in this section) through forward projection. However, problems often arise in the following scenarios:

1. Parts of the object in 3D are occluded by unwanted objects (e.g. trees occluding a building façade).
2. Images from multiple views vary in lighting conditions and distance to the object of interest, leading to unbalanced colour and inconsistent resolution.
3. Multi-view overlapping images result in repetitive coverage of the textures, How to select the best images for cropping the textures.

The first problem is related to object detection and texture synthesis, as the unwanted objects need to be detected and then the occluded area be replaced by synthetic textures (Bincy, 2012; Böhm, 2004). In practice (engineering-grade data generation), this is often done by manual editing, or by introducing more convergence images covering the objects. Most of the state-of-the-art texture mapping solutions tend to address problems (2) and (3). Given oriented images and triangular meshes/polyhedral models, a texture mapping procedure aims to find for each triangle/face, an optimal texture patch directly from a single image, or a fused/blended image of an optimally selected set of images. It often contains three steps:

1. *Visibility analysis.* Test if the faces are occluded by other objects, visible or partially visible in the ray direction of the considered images, such that a list of candidate images can be identified. Details are described in Section 3.5.1.
2. *Image assignment and colour blending.* Based on the candidate images, the best set of images is selected, and colours from these images are blended in the faces. Details are described in Section 3.5.2.
3. *Seam elimination.* Adjacent texture patches may come from different image sets, and this creates seams in their boundaries. This step aims to eliminate these seams. See Section 3.5.3.

### 3.5.1 Visibility analysis

The visibility of a triangle/face is associated with image views. A common approach to test if a face is visible in a view is to compare its distance to the image perspective center with those of others. A face with all its points closest to the perspective center of the view is deemed visible, or, if part of its points are visible, being partially visible, this face is deemed occluded. This is essentially the well-known Z-buffer algorithm (Bernardini *et al.*, 2001; Nguyen *et al.*, 2013; Velho and Sossai, 2007), as shown in Figure 3.10.

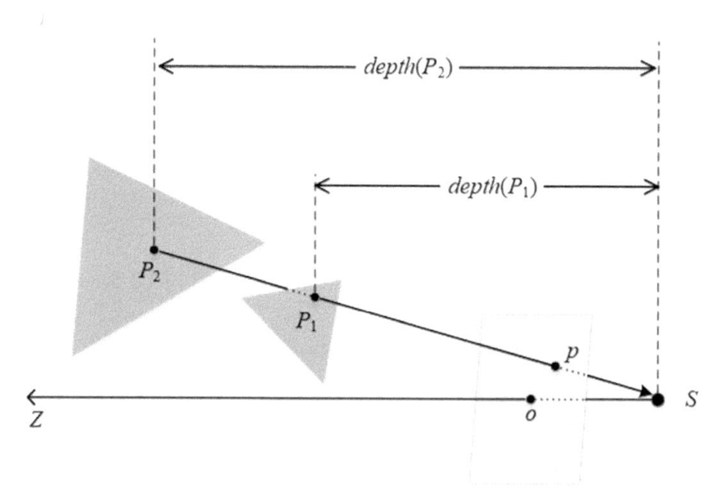


Figure 3.10 Projection of meshes in Z-buffer ($P_1$ being visible). $S$ represents the camera center; $o$ represents the principal point in the image plane; $p$ represents the projective pixel; $\overrightarrow{SZ}$ represents the principal optic axis; $P_1$, $P_2$ represent 3D points in faces corresponding to the same image pixel.

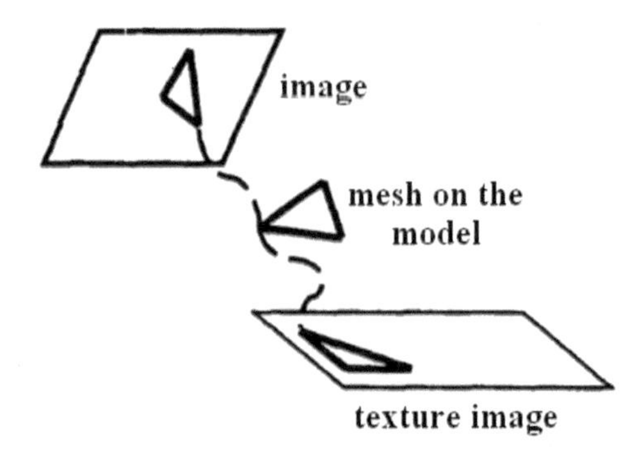


Figure 3.11 Texture images for mapping (Li *et al.*, 2010).

In the practical implementation, testing over all the images for each face is computationally intensive, while a simple intersection angle test is very effective: images with their central-ray forming an intersection angle with the face normal greater than 90° can be discarded, as their contribution towards the texture mapping is insignificant (Li *et al.*, 2010). By determining eligible image patches to be mapped on the face, two straightforward processes need to be implemented: (1) crop out the face footprint in the image that is beyond the image dimension, and (2) invalidate occluded areas in the face footprint. The image patches in the remaining face footprint area are eligible for mapping onto the faces.

It is a slightly complicated problem to map partial textures to a face, as the mapping is performed based on the vertices. One solution is to search for other images where the face is fully contained. This is usually reasonable for triangular meshes as they are typically small in units and the chances of finding images where one triangle is visible are high. However, for larger polyhedral faces, this is usually solved by re-synthesizing a new texture patch from difference images that cover the entire image (UV mapping method, texture image shown in Figure 3.11) (Desbrun *et al.*, 2002; Levy *et al.*, 2002; Maillot *et al.*, 1993; Zigelman *et al.*, 2002).

### 3.5.2 Image assignment and colour blending

The candidate images corresponding to a face may vary in geometric accuracy, radiance, resolution and their intersection angles to the face, etc. A simple image averaging usually creates blurring, ghosting effects (Bernardini *et al.*, 2001). There are generally two classes of method for image fusion and blending: (1) *Weighted colour blending* (Bernardini *et al.*, 2001; Callieri *et al.*, 2008; Grammatikopoulos *et al.*, 2007; Li *et al.*, 2010); and (2) *Best image selection* (Goldberg, 2014; Hanusch, 2009; Lempitsky, 2007; Niem and Broszio, 1995).

*Weighted colour blending*: This method weights the texture patches from different images/views and blends these images by linearly combining the weighted image colours. The commonly used factors for determining the weights include *image distance, intersection angle, texture patch size, colour difference, patch position in the original image*. These factors and their combinations are used in different methods, and some of them are correlated: e.g. *image distance* and *texture patch size* are both indicators for the texture resolution. In addition, the weighting scheme can be applied for either each patch or each ray (pixel). *Imaging distance* refers to the distance from the image perspective center to the 3D face, negatively correlated with resolution; and *texture patch size* refers to the size of the face footprint mapped onto the image, positively correlated with resolution. And usually the higher the resolution, the larger the weighting one image will be assigned.

*Intersection angle* is related to the distortion of the texture on the face. The smaller the intersection angle, the smaller the distortion of the texture on the face, the larger the weighting that will be assigned. *Patch position in the original image* is an easier representation of the intersection angle. It measures the position of the face footprint to the center of the image: the further it is, the larger the intersection angle, the smaller the weighting to be assigned. The *colour difference* measures the difference in the pixel colours to the median or mean image over all the candidate texture patches. The larger difference the is, the smaller the weightings are. This assumes that the colour distribution of a pixel should have a mean center, where images with large differences should only be given a small weighting.

*Best image selection*: Sometimes only one best image is sufficient rather than blending a number of candidate texture patches (which may introduce errors). The selection of this single best image takes into account similar indicators as the weighted colour blending method, and can be carried out either per face (local selection) (Goldberg, 2014; Hanusch, 2009), or by leveraging the colour balancing across all the faces (global selection) (see Figure 3.12). Local selection optimizes a single indicator combined with various indicators as used in weighted colour blending (*image distance, intersection angle, texture patch size, colour difference, patch position in the original image*). This normally generates seams between faces. The global method imposes consistencies of image selection over adjacent faces (Lempitsky, 2007; Niem and Broszio, 1995). This can be formulated as an energy minimization problem (equation (5)). Being a multi-labeling problem this can be solved via classic algorithms, e.g. graph cut (Lempitsky, 2007) or the two-steps method (Niem and Broszio, 1995).

### 3.5.3 Seam elimination

Seams between adjacent faces, in faces with partial textures combined (also reflecting corresponding parts in the texture image), are unavoidable even when they have undergone the blending procedure (in section 3.5.2). An example is shown in Figure 3.13.

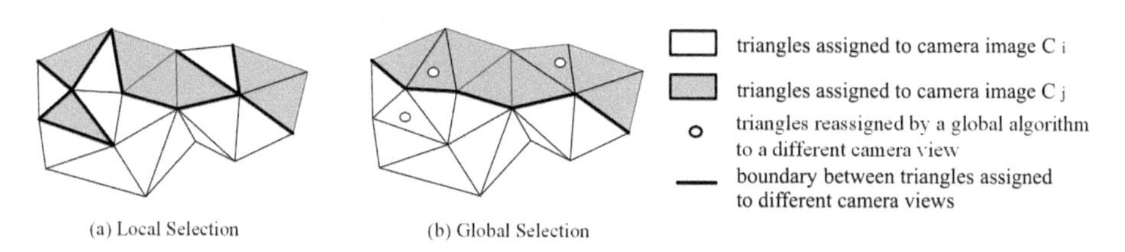


Figure 3.12 Image selection after Global Selection (Niem and Broszio, 1995).

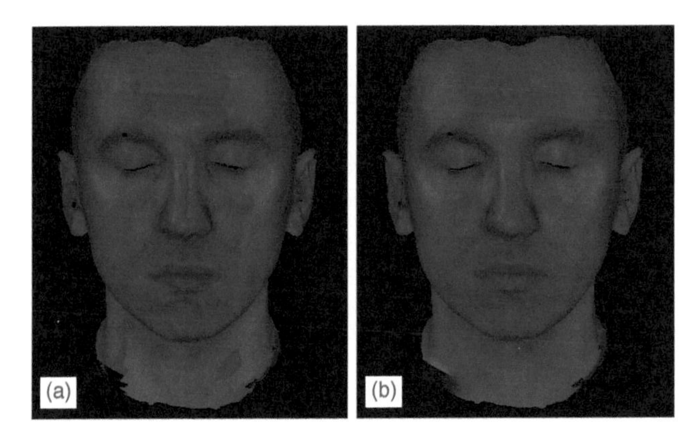


Figure 3.13 Before and after seam elimination (Velho and Sossai, 2007): (a) texture mapping result before seam elimination; (b) texture mapping result after elimination.

Such seams not only affect the visual appearance but they also create challenges for 3D model interpretation. We introduce two representative methods that deal with seams in texture mapping.

### Levelling function based correction

Considering the texture space as a function, seams appears as colour discontinuities between meshes or fragments. The leveling function approach aims to construct an auxiliary function that compensates for the discontinuities in the texture surfaces (Lempitsky, 2007; Niem and Broszio, 1995; Velho and Sossai, 2007). The idea of leveling functions is to compensate large colour jumps by reverting the gradients in these jumps and constructing linear functions between these jumps. By adding this auxiliary function to the original function (texture colours), discontinuities can be leveled up. Figure 3.14 gives an example of this idea. The leveling function (Figure 3.14b) retains the large jumps (but with a reverse direction) while keeping the other corresponding part smooth.

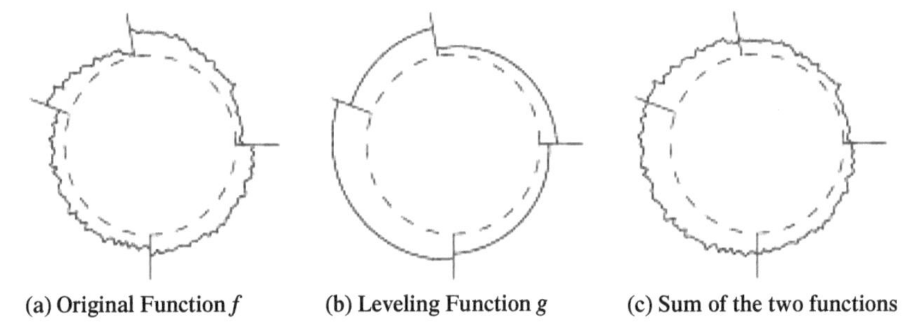


Figure 3.14 Seam leveling on a circumference (Lempitsky, 2007). The function values are shown as the elevations of the circumference.

*Brightness correction*: Brightness correction assumes that brightness difference is the main origin of seams. Brightness values can be computed as the lightness (L) component through colour space transformation (*Hue, Saturation, and Lightness* (HSL) or *Colour Space Adopted in International Commission on Illumination* (CIELAB)) (Zhou *et al.*, 2014). The essential of brightness correction is to minimize the differences between the two L components of adjacent faces, and correct the relative brightness values. The simplest way is to compute a constant difference between the two L components and compensate. However, the brightness difference may become complicated and errors for such correction might be propagated. Therefore, this method is usually used for local correction (Hanusch, 2009).

## 3.6 Summary

This chapter provides an overview of the geometric processing of image-based RBM modelling. In particular, this includes three major topics: (1) image-based georeferencing; (2) dense image matching; and (3) texture mapping. Each of these topics has been heavily investigated by scientists in the computer vision and photogrammetry community. Although geometric processing techniques can be largely automated nowadays, particularly for regularly acquired images, our review of methods yet reveals the fact that the components that require intelligence – for example, feature point extraction, dense corresponding search, and unwanted object removal in texture mapping – are still somehow data and scene dependent. The widely used piecewise linear assumption is not sufficient to cover all types of complex objects in reality, and this usually fails for fence-like objects or tree structures. Moreover, methods that in practice work on critical scenarios, such as reflecting surfaces and transparent surface, are still lacking. Across the field of processing, there is an overall lack of learning-based methods that account for such critical situations (e.g. occluded objects in texture mapping). With the prevalent machine learning techniques applied in various domains in computer vision and photogrammetry, we foresee more and more effort using machine learning approaches will be devoted to solving these problems.

## References

Agarwal, S., Y. Furukawa, N. Snavely, I. Simon, B. Curless, S. M. Seitz and R. Szeliski (2011) Building Rome in a day. *Communications of the ACM* **54**(10), 105–112.

Agisoft (2017) Photoscan, http://www.agisoft.com/. (last date accessed: 9 June 2017).

Andreasson, H., R. Triebel and A. J. Lilienthal (2006) Vision-based interpolation of 3D laser scans. In: *Proceedings of International Conference on Autonomous Robots and Agents (ICARA)*, Palmerston North, New Zealand, 83–90.

Bao, S. Y., M. Chandraker, Y. Lin & S. Savarese (2013) Dense object reconstruction with semantic priors. In: *Proceedings/CVPR, IEEE Computer Society Conference on Computer Vision and Pattern Recognition*, 1264–1271.

Barnes, C., E. Shechtman, A. Finkelstein and D. B. Goldman (2009) Patchmatch: A randomized correspondence algorithm for structural image editing. *ACM Transactions on Graphics* **28**(3), 341–352.

Bay, H., T. Tuytelaars and L. Van Gool (2006) Surf: Speeded up robust features. In: *Computer vision–ECCV* 2006, 404–417.

Bernardini, F., I. Martin and H. Rushmeier (2001) High-quality texture reconstruction from multiple scans. *IEEE Transactions on Visualization and Computer Graphics* **7**(4), 318–332.

Biljecki, F., J. Stoter, H. Ledoux, S. Zlatanova and A. Çöltekin (2015) Applications of 3D city models: State of the art review. *ISPRS International Journal of Geo-Information* **4**(4), 2842–2889.

Bincy, A. M. (2012) Removing occlusion in images using sparse processing and texture synthesis. *International Journal of Computer Science Engineering & Applications* **2**(3), 117–124.

Birchfield, S. and C. Tomasi (1998) A pixel dissimilarity measure that is insensitive to image sampling. *IEEE Transactions on Pattern Analysis and Machine Intelligence* **20**(4), 401–406.

Blaha, M., C. Vogel, A. Richard, J. D. Wegner, T. Pock and K. Schindler (2016) Large-scale semantic 3D reconstruction: An adaptive multi-resolution model for multi-class volumetric labeling. In: *IEEE Conference on Computer Vision and Pattern Recognition (CVPR)*, 3176–3184.

Bleyer, M. and M. Gelautz (2005) A layered stereo matching algorithm using image segmentation and global visibility constraints. *ISPRS Journal of Photogrammetry and Remote Sensing* **59**(3), 128–150.

Böhm, J. (2004) Multi-image fusion for occlusion-free façade texturing. *Rendering Techniques* **35**(5), 867–872.

Boykov, Y. and V.S. Lempitsky (2013) From photohulls to photoflux optimization. In: *British Machine Vision Conference 2006*, Edinburgh, UK, 1149–1158.

Broadhurst, A., T.W. Drummond and R. Cipolla (2001) A probabilistic framework for space carving. In: *Proceedings of IEEE International Conference of Computer Vision*, 388–393.

Brown, D. C. (1976) The bundle adjustment – Progress and prospects. *Int. Archives Photogrammetry* **21**(3), Paper 3-03.

Callieri, M., P. Cignoni, M. Corsini and R. Scopigno (2008) Masked photo blending: Mapping dense photographic dataset on high-resolution sampled 3D models. *Computers & Graphics* **32**(4), 464–473.

Cornelis, N., B. Leibe, K. Cornelis and L. Van Gool (2008) 3D urban scene modelling integrating recognition and reconstruction. *International Journal of Computer Vision* **78**(2–3), 121–141.

D'Angelo, P. (2013) *High Quality DSM Generation*. Technical Report. Deutsches Zentrum für Luft- und Raumfahrt (DLR).

Desbrun, M., M. Meyer and P. Alliez (2002) Intrinsic parameterizations of surface meshes. *Computer Graphics Forum* **21**(3), 209–218.

Deseilligny, M. P. and I. Clery (2011) Apero, an open source bundle adjusment software for automatic calibration and orientation of set of images. In: *Proceedings of ISPRS International Workshop on 3D Virtual Reconstruction and Visualization of Complex Architectures*, Trento, Italy, 2–4 March, 269–276.

Diakité, A. A., G. Damiand and D. Van Maercke (2014) Topological reconstruction of complex 3D buildings and automatic extraction of levels of detail. In: *Eurographics Workshop on Urban Data Modelling and Visualisation*, April, Strasbourg, 25–30.

Diebel, J. and S. Thrun (2005) An application of Markov random fields to range sensing. In: *Advances in Neural Information Processing Systems 18*, eds Weiss, Y., Scholkopf, B.and Plat, J. Cambridge, MA: MIT Press, 291–298.

Doyle, F. (1964) The historical development of analytical photogrammetry. *Photogrammetric Engineering* **30**(2), 259–265.

Drouyer, S., S. Beucher, M. Bilodeau, M. Moreaud and L. Sorbier (2017) Sparse stereo disparity map densification using hierarchical image segmentation. In: *International Symposium on Mathematical Morphology and Its Applications to Signal and Image Processing*, 172–184.

Duan, Y., X. Huang, J. Xiong, Y. Zhang and B. Wang (2016) A combined image matching method for Chinese optical satellite imagery. *International Journal of Digital Earth* **9**(9), 851–872.

Faugeras, O. and R. Keriven (1998) Variational principles, surface evolution, PDE's, level set methods and the stereo problem. *IEEE Transactions on Image Processing* **7**(3), 336–344.

Ferretti, A., C. Prati and F. Rocca (2001) Permanent scatterers in SAR interferometry. *IEEE Transactions on Geoscience and Remote Sensing* **39**(1), 8–20.

Flamanc, D., G. Maillet and H. Jibrini (2003) 3D city models: An operational approach using aerial images and cadastral maps. In: *International archives of photogrammetry remote sensing and spatial information sciences* **34**(3/W8), 53–58.

Förstner, W., T. Dickscheid and F. Schindler (2009) Detecting interpretable and accurate scale-invariant keypoints. In: *2009 IEEE 12th International Conference on Computer Vision*, 2256–2263.

Förstner, W. and E. Gülch (1987) A fast operator for detection and precise location of distinct points, corners and centres of circular features. In: *Proc. ISPRS intercommission conference on Fast Processing of Photogrammetric Data*, 2–4 June, Interlaken, 281–305.

Frahm, J.-M., P. Fite-Georgel, D. Gallup, T. Johnson, R. Raguram, C. Wu, Y.-H. Jen, E. Dunn, B. Clipp and S. Lazebnik (2010) Building Rome on a cloudless day. In: *European Conference on Computer Vision*, 368–381.

Fraser, C. S. (2013) Automatic camera calibration in close range photogrammetry. *Photogrammetric Engineering & Remote Sensing* **79**(4), 381–388.

Fraundorfer, F., P. Tanskanen and M. Pollefeys (2010) A minimal case solution to the calibrated relative pose problem for the case of two known orientation angles. In: *Computer Vision–ECCV* 2010, 269–282.

Freitas, S., C. Catita, P. Redweik and M. C. Brito (2015) Modelling solar potential in the urban environment: State-of-the-art review. *Renewable and Sustainable Energy Reviews* **41**, 915–931.

Furukawa, Y., B. Curless, S. M. Seitz & R. Szeliski (2010) Towards internet-scale multi-view stereo. In: *Conference on Computer Vision and Pattern Recognition*, 1434–1441

Furukawa, Y. and J. Ponce (2007) Accurate, dense, and robust multi-view stereopsis. *IEEE Trans. Pattern Anal.*. 32. 10.1109/CVPR.2007.383246.

Gehrke, S., K. Morin, M. Downey, N. Boehrer and T. Fuchs (2010) Semi-global matching: An alternative to LIDAR for DSM generation. In: *International Archives of the Photogrammetry, Remote Sensing and Spatial Information Sciences*, Calgary, AB, **38**(B1) 6.

Gherardi, R., M. Farenzena and A. Fusiello (2010) Improving the efficiency of hierarchical structure-and-motion. In: *IEEE Conference on Computer Vision and Pattern Recognition (CVPR)*, 2010, 1594–1600.

Goldberg, D., C. Salvaggio and C. F. Carlson (2014) Seamless texture mapping of 3D point clouds, https://pdfs.semanticscholar.org/8c8a/21d7b0681245182fbaa8f2001bf71ff4c830.pdf (accessed 25 November 2014).

Graça, N., E. Mitishita and J. Gonçalves (2014) Photogrammetric mapping using unmanned aerial vehicles. In: *The International Archives of the Photogrammetry, Remote Sensing and Spatial Information Sciences*, 129–133.

Grammatikopoulos, L., I. Kalisperakis, G. Karras and E. Petsa (2007) Automatic multi-view texture mapping of 3D surface projections. In: *3D Virtual Reconstruction & Visualization of Complex Architectures*, 12–13.

Gruen, A. (1985) Adaptive least squares correlation: A powerful image matching technique. *South African Journal of Photogrammetry, Remote Sensing and Cartography* **14**(3), 175–187.

Gruen, A. (1998) TOBAGO – A semi-automated approach for the generation of 3-D building models. *ISPRS Journal of Photogrammetry and Remote Sensing* **53**(2), 108–118.

Gruen, A. (2008) Reality-based generation of virtual environments for digital earth. *International Journal of Digital Earth* **1**(1), 88–106.

Gruen, A. (2013) Next generation smart cities – The role of geomatics. *International Workshop on Global Geospatial Information*, Siberian State Academy of Geodesy, 25–41.

Gruen, A., M. Behnisch and N. Kohler (2009) Perspectives in the reality-based generation, n D modelling, and operation of buildings and building stocks. *Building Research & Information* **37**(5–6), 503–519.

Gruen, A. and H. A. Beyer (2001) System calibration through self-calibration. In: *XVII ISPRS Congress*, Washington, 215–225.

Gruen, A., X. Huang, R. Qin, T. Du, W. Fang, J. Boavida and A. Oliveira (2013) Joint processing of UAV imagery and terrestrial mobile mapping system data for very high resolution city modelling. *ISPRS-International Archives of the Photogrammetry, Remote Sensing and Spatial Information Sciences* **1**(2), 175–182.

Gruen, A. and X. Wang (1998) CC-Modeler: A topology generator for 3-D city models. *ISPRS Journal of Photogrammetry and Remote Sensing* **53**(5), 286–295.

Grün, A. (1985) Algorithmic aspects of on-line triangulation. *Photogrammetric Engineering & Remote Sensing* **4**(51), 419–436.

Guney, F. and A. Geiger (2015) Displets: Resolving stereo ambiguities using object knowledge. In: *IEEE Conference on Computer Vision and Pattern Recognition*, 2015, Boston, 4165–4175.

Haala, N. and M. Kada (2010) An update on automatic 3D building reconstruction. *ISPRS Journal of Photogrammetry and Remote Sensing* **65**(6), 570–580.

Habib, A. F., M. Morgan and Y. R. Lee (2002) Bundle adjustment with self–calibration using straight lines. *The Photogrammetric Record* **17**(100), 635–650.

Häne, C., C. Zach, A. Cohen and M. Pollefeys (2016) Dense semantic 3D reconstruction. *IEEE Transactions on Pattern Analysis and Machine Intelligence* **14**(8), 1–14.

Hanusch, T. (2009) A new texture mapping algorithm for photorealistic reconstruction of 3D objects. In: *ISPRS Congress*, Commission V, WG V/4, 699–706.

Hanusch, T. (2010) Texture mapping and true orthophoto generation of 3D objects. *USSR Computational Mathematics & Mathematical Physics* **14**(2), 208–212.

Harris, C. and M. Stephens (1988) A combined corner and edge detector. In: *Proceedings of the Fourth Alvey Vision Conference*, 147–151.

Hartley, R. and A. Zisserman (2004) *Multiple View Geometry in Computer Vision*. Cambridge: Cambridge University Press, 672pp.

Havlena, M. and K. Schindler (2014) Vocmatch: Efficient multiview correspondence for structure from motion. In: *European Conference on Computer Vision*, 46–60.

He, K., J. Sun. and X. Tang (2013) Guided image filtering. *IEEE Transactions on Pattern Analysis and Machine Intelligence* **35**(6), 1397–1409.

Hirschmüller, H. (2008) Stereo processing by semiglobal matching and mutual information. *IEEE Transactions on Pattern Analysis And Machine Intelligence* **30**(2), 328–341.

Hirschmüller, H. and D. Scharstein (2009) Evaluation of stereo matching costs on images with radiometric differences. *IEEE Transactions on Pattern Analysis and Machine Intelligence* **31**(9), 1582–1599.

Huang, X. (2013) Building reconstruction from airborne laser scanning data. *Geo-spatial Information Science* **16**(1), 35–44.

Huang, X., K. Hu, X., Ling, Y., Zhang, Z. Lu and G. Zhou. (2017) Global patch matching. In: *ISPRS Annals of the Photogrammetry, Remote Sensing and Spatial Information Sciences, Volume IV-2/W4, 2017*

Huang, X., Y. Zhang and Z. Yue (2016) Image-guided non-local dense matching with three-steps optimization. In: *ISPRS Annals of Photogrammetry, Remote Sensing and Spatial Information Sciences*, Prague, 67–74.

*ISPRS Geospatial Week*, Wuhan, 227–234.

Jiao, J. B., R. G. Wang, W. M. Wang and S. Dong (2014) Local stereo matching with improved matching cost and disparity refinement. *IEEE Multimedia* **21**(4), 16–27.

Kanade, T. and M. Okutomi (1994) A stereo matching algorithm with an adaptive window: Theory and experiment. *IEEE Transactions on Pattern Analysis and Machine Intelligence* **16**(9), 920–932.

Ke, Y. and R. Sukthankar (2004) PCA-SIFT: A more distinctive representation for local image descriptors. In: *Conference on Computer Vision and Pattern Recognition, 2004. CVPR 2004. Proceedings of the 2004 IEEE Computer Society*, 506–513.

Kim, K. R. and C. S. Kim (2016) Adaptive smoothness constraints for efficient stereo matching using texture and edge information. In: *International Conference on Image Processing, Phoenix*, 3429–3434.

Klaus, A., M. Sormann and K. Karner (2006) Segment-based stereo matching using belief propagation and a self-adapting dissimilarity measure. In: *International Conference on Pattern Recognition, Hong Kong*, 15–18.

Koch, R., M. Pollefeys and L. Van Gool (1998) Multi viewpoint stereo from uncalibrated video sequences. In: *Computer vision–ECCV*, 55–71.

Kolmogorov, V. and R. Zabih (2001) Computing visual correspondence with occlusions using graph cuts. In: *International Conference on Computer Vision, Vancouver, BC, Canada*, 508–515.

Kolmogorov, V. and R. Zabih (2002) Multi-camera scene reconstruction via graph cuts. In: *Proceedings of the European Conference on Computer Vision, Copenhagen, Denmark*, 28–31.

Kordelas, G. A., D. S. Alexiadis and P. Daras (2015) Enhanced disparity estimation in stereo images. *Image and Vision Computing* **35**, 31–49.

Krizhevsky, A., I. Sutskever and G. E. Hinton (2012) Imagenet classification with deep convolutional neural networks. In: *Advances in Neural Information Processing Systems*, eds Weiss, Y., Scholkopf, B.and Plat, J. Cambridge, MA: MIT Press, 1097–1105.

Kutulakos, K. N. and S. M. Seitz (2002) A theory of shape by space carving. *International Journal of Computer Vision* **38**(3), 199–218.

Ladický, L., P. Sturgess, C. Russell, S. Sengupta, Y. Bastanlar, W. Clocksin and P. H. S. Torr, (2012) Joint optimisation for object class segmentation and dense stereo reconstruction. *International Journal of Computer Vision* **100**(2), 122–133.

LeCun, Y., Y. Bengio and G. Hinton (2015) Deep learning. *Nature* **521**(7553), 436–444.

Lempitsky, V. (2007) Seamless mosaicing of image-based texture maps. In: *IEEE Conference on Computer Vision and Pattern Recognition*, pp. 1–6.

Lempitsky, V., Y. Boykov and D. Ivanov (2006) Oriented visibility for multiview reconstruction. In: *Proceedings of European Conference on Computer Vision*, 226–238.

Levy, B., S. Petitjean, N. Ray and J. Maillot (2002) Least squares conformal maps for automatic texture atlas generation. *ACM Transactions on Graphics* **21**(3), 362–371.

Lewis, J. P. (1995) Fast normalized cross-correlation. In: *Proceedings of Vision Interface*, 120–123.

Li, J., Z. Miao, X. Liu and Y. Wan (2010) 3D reconstruction based on stereovision and texture mapping. In: Paparoditis N., Pierrot-Deseilligny M., Mallet C., Tournaire O. (eds), *IAPRS, Vol. XXXVIII, Part 3B, Saint-Mande, France*, 1–6.

Li, L., S. Zhang, X. Yu and L. Zhang (2016) PMSC: PatchMatch-based superpixel cut for accurate stereo matching. *IEEE Transactions on Circuits and Systems for Video Technology* (99), 1–14.

Li, X., C. Wu, C. Zach, S. Lazebnik and J.-M. Frahm (2008) Modelling and recognition of landmark image collections using iconic scene graphs. In: *European Conference on Computer Vision, ECCV 2008*, 427–440.

Liebelt, J. and C. Schmid (2010) Multi-view object class detection with a 3d geometric model. In: *IEEE Conference on Computer Vision and Pattern Recognition (CVPR)*, 2010, 1688–1695.

Lowe, D. G. (2004) Distinctive image features from scale-invariant keypoints. *International Journal of Computer Vision* **60**(2), 91–110.

Maillot, J., H. Yahia and A. Verroust (1993) Interactive texture mapping. In: *Proceedings of the 20th Annual Conference on Computer Graphics and Interactive Techniques, SIGGRAPH* 1993, 27–34.

Matusik, W., C. Buehler, R. Raskar, S. J. Gortler and L. Mcmillan (2000) Image-based visual hulls. In: *Proceedings of the 27th annual conference on Computer Graphics and Interactive Techniques SIGGRAPH 2000*, 369–374.

Mikhail, E. M., J. S. Bethel and J. C. McGlone (2001) *Introduction to Modern Photogrammetry.* New York: Wiley.

Moravec, H. P. (1980) Obstacle avoidance and navigation in the real world by a seeing robot rover. DTIC Document, Stanford University.

Morel, J.-M. and G. Yu (2009) ASIFT: A new framework for fully affine invariant image comparison. *SIAM Journal on Imaging Sciences* **2**(2), 438–469.

Mozerov, M. G. and J. van de Weijer (2015) Accurate stereo matching by two-step energy minimization. *IEEE Transactions on Image Processing* **24**(3), 1153–1163.

Mundy, J. (1993) The relationship between photogrammetry and computer vision. In: *Proc. SPIE*, 92–105.

Newcombe, R., S. Izadi, O. Hilliges, D. Molyneaux, D. Kim, A.J. Davison, P. Kohi, J. Shotton, S. Hodges and A. Fitzgibbon (2011) KinectFusion: Real-time dense surface mapping and tracking. In: *IEEE International Symposium on Mixed and Augmented Reality*, 127–136.

Nex, F. and F. Remondino (2014) UAV for 3D mapping applications: A review. *Applied Geomatics* **6**(1), 1–15.

Nguyen, H. M., B. Wünsche and P. Delmas (2013) High-definition texture reconstruction for 3D image-based modelling. In: *21st International Conference on Computer Graphics, Visualization and Computer Vision*, 39–48.

Niederöst, M. (2003) *Detection and Reconstruction of Buildings for Automated Map Updating.* Zurich: Mitteilungen- Institut fur Geodasie und Photogrammetrie an der Eidgenossischen Technischen Hochschule Zurich.

Niem, W. and H. Broszio (1995) Mapping texture from multiple camera views onto 3D-object models for computer animation. In: *Proceedings of the International Workshop on Stereoscopic & Three Dimensional Imaging*, 99–105.

Nistér, D., O. Naroditsky and J. Bergen (2004) Visual odometry. In: *Proceedings of the 2004 IEEE Computer Society Conference on Computer Vision and Pattern Recognition, CVPR 2004*, 652–659.

Nister, D. and H. Stewenius (2006) Scalable recognition with a vocabulary tree. In: *Proceedings of the 2006 IEEE Computer Society Conference on Computer Vision and Pattern Recognition*, 2161–2168.

Olsson, C., J. Ulen and Y. Boykov (2013) In defense of 3D-label stereo. In: *Proceedings of the 2013 IEEE Computer Society Conference on Computer Vision and Pattern Recognition*, Portland, 1730–1737.

Papadakis, N. and V. Caselles (2010) Multi-label depth estimation for graph cuts stereo problems. *Journal of Mathematical Imaging & Vision* **38**(1), 70–82.

Paul, V. and W. M. Wells III (1997) Alignment by maximization of mutual information. *International Journal of Computer Vision* **24**(2), 137–154.

Pham, C. C. and J. W. Jeon (2013) Domain transformation-based efficient cost aggregation for local stereo matching. *IEEE Transactions on Circuits and Systems for Video Technology* **23**(7), 1119–1130.

Pix4D. (2017) Pix4D, http://pix4d.com/ (last accessed: 9 June 2017).

Pons, J. P., R. Keriven and O. Faugeras (2007) Multi-view stereo reconstruction and scene flow estimation with a global image-based matching score. *International Journal of Computer Vision* **72**(2), 179–193.

Qin, R. (2014) An object-based hierarchical method for change detection using unmanned aerial vehicle images. *Remote Sensing* **6**(9), 7911–7932.

Qin, R., A. Gruen and X. Huang (2012) UAV project – Building a reality-based 3D model of the NUS (National University of Singapore) Campus. In: *Asian Conference of Remote Sensing, Pattaya, Thailand, November*, 26–30, E1-4.

Remondino, F. (2006) Detectors and descriptors for photogrammetric applications. *International Archives of Photogrammetry, Remote Sensing and Spatial Information Sciences* **36**(3), 49–54.

Remondino, F. and C. Fraser (2006) Digital camera calibration methods: Considerations and comparisons. *International Archives of Photogrammetry, Remote Sensing and Spatial Information Sciences* **36**(5), 266–272.

Remondino, F., M. G. Spera, E. Nocerino, F. Menna and F. Nex (2014) State of the art in high density image matching. *The Photogrammetric Record* **29**(146), 144–166.

Rhemann, C., A. Hosni, M. Bleyer, C. Rother and M. Gelautz (2011) Fast cost-volume filtering for visual correspondence and beyond. In: *Proceedings of the 2011 IEEE Computer Society Conference on Computer Vision and Pattern Recognition, CVPR 2011*, 3017–3024.

Sato, T., M. Kanbara, N. Yokoya and H. Takemura (2002) Dense 3D reconstruction of an outdoor scene by hundreds-baseline stereo using a hand-held video camera. *International Journal of Computer Vision* **47**(1–3), 119–129.

Savinov, N., C. Haene, L. Ladicky and M. Pollefeys (2016) Semantic 3D reconstruction with continuous regularization and ray potentials using a visibility consistency constraint. In: *Proceedings of the 2011 IEEE Computer Society Conference on Computer Vision and Pattern Recognition*, 5460–5469.

Scharstein, D. and R. Szeliski (2002) A taxonomy and evaluation of dense two-frame stereo correspondence algorithms. *International Journal of Computer Vision* **47**(1–3), 7–42.

Scharstein, D. and R. Szeliski (2014) Middlebury stereo vision page, http://vision.middlebury.edu/stereo/ (accessed 3 December, 2014).

Schenk, T. (2005) Introduction to photogrammetry. *The Ohio State University, Columbus* **106**

Schwarz, B. (2010) LIDAR: Mapping the world in 3D. *Nature Photonics* **4**(7), 429.

Shackelford, A. K. and C. H. Davis (2003) A combined fuzzy pixel-based and object-based approach for classification of high-resolution multispectral data over urban areas. *IEEE Transactions on Geoscience and Remote Sensing* **41**(10), 2354–2363.

Slabaugh, G. and G. Unal (2005) Active polyhedron: Surface evolution theory applied to deformable meshes. In: *IEEE Computer Society Conference on Computer Vision and Pattern Recognition*, 84–91.

Snavely, N. (2010) Bundler: Structure from motion (SFM) for unordered image collections. Available online: phototour. cs. washington. edu/bundler/ (accessed on 12 July 2013).

Sohn, G., X. Huang and V. Tao (2008) Using a binary space partitioning tree for reconstructing polyhedral building models from airborne lidar data. *Photogrammetric Engineering & Remote Sensing* **74**(11), 1425–1438.

Strzalka, A., J. Bogdahn, V. Coors and U. Eicker (2011) 3D City modelling for urban scale heating energy demand forecasting. *HVAC&R Research* **17**(4), 526–539.

Taniai, T., Y. Matsushita and T. Naemura (2014) Graph cut based continuous stereo matching using locally shared labels. In: *IEEE Conference on Computer Vision and Pattern Recognition, Columbus, OH, USA*, 1613–1620.

Tao, P. (2016) 3D Surface reconstruction and optimization based on geometric and radiometric integral imaging model. PhD thesis, School of Remote Sensing and Information Engineering, Wuhan University, China.

Turk, G. and M. Levoy (1994) Zippered polygon meshes from range images. In: *Proceedings of the 21st Annual Conference on Computer Graphics and Interactive Techniques, SIGGRAPH 1994,* 311–318.

Tyleček, R. and R. Šára (2010) Refinement of surface mesh for accurate multi-view reconstruction. *The International Journal of Virtual Reality*, **9**(1), 45–54.

Veldandi, M., S. Ukil and K.A. Govindarao (2014) Robust segment-based stereo using cost aggregation. In: *Proceedings of the British Machine Vision Conference*, Nottingham, 1–11.

Velho, L. and J. Sossai (2007) Projective texture atlas construction for 3D photography. *The Visual Computer* **23**(9), 621–329.

Verdie, Y., F. Lafarge and P. Alliez (2015) LOD generation for urban scenes. *ACM Transactions on Graphics* **34**(3), 1–14.

Vernay, D. G., B. Raphael and I. F. Smith (2015) A model-based data-interpretation framework for improving wind predictions around buildings. *Journal of Wind Engineering and Industrial Aerodynamics* **145**, 219–228.

Vogiatzis, G., C. Hernandez, P. H. Torr and R. Cipolla (2007) Multiview stereo via volumetric graph-cuts and occlusion robust photo-consistency. *IEEE Transactions on Pattern Analysis and Machine Intelligence* **29**(12), 2241–2246.

Vu, H., P. Labatut, J. Pons and R. Keriven (2012) High accuracy and visibility-consistent dense multiview stereo. *IEEE Transactions on Pattern Analysis and Machine Intelligence* **34**(5), 889–901.

Wang, R. and F. P. Ferrie (2015) Upsampling method for sparse light detection and ranging using coregistered panoramic images. *Journal of Applied Remote Sensing* **9**(1), 095075-1-15.

Wei, D., C. Liu and W. T. Freeman (2014) A data-driven regularization model for stereo and flow. In: International Conference on 3d Vision, pp. 277–284.

Wenzel, K., M. Rothermel, N. Haala and D. Fritsh (2013) SURE – The ifp Software for Dense Image Matching. In: Photogrammetric Week '13, Belin, pp. 59–70.

Wu, C. (2013) Towards linear-time incremental structure from motion. In: 3DTV-Conference, 2013 International Conference on, pp. 127–134.

Wu, C. (2014) VisualSFM : A visual structure from motion system, http://ccwu.me/vsfm/ (last accessed: 26 February 2014).

Wu, C., S. Agarwal, B. Curless and S. Seitz (2011) Multicore bundle adjustment. In: *CVPR '11: Proceedings of the 2011 IEEE Conference on Computer Vision and Pattern Recognition*, pp. 3057–3064.

Xiong, B., S. O. Elberink and G. Vosselman (2014) A graph edit dictionary for correcting errors in roof topology graphs reconstructed from point clouds. *ISPRS Journal of Photogrammetry and Remote Sensing* **93**, 227–242.

Yamaguchi, K., T. Hazan., D. Mcallester & R. Urtasun (2012) Continuous Markov random fields for robust stereo estimation. In: Fitzgibbon, A., Lazebnik, S., Perona, P., Sato, Y., Schmid, C. (eds) Computer Vision – ECCV 2012. ECCV 2012. Lecture Notes in Computer Science, vol. 7576. Springer, Berlin, Heidelberg, 45–58.

Yamaguchi, K., D. Mcallester and R. Urtasun. (2014) Efficient joint segmentation, occlusion labeling, stereo and flow estimation. In: European Conference on Computer Vision, pp. 756–771.

Yang, Q.X. (2015) Stereo matching using tree filtering. *IEEE Transactions on Pattern Analysis and Machine Intelligence* **37**(4), pp. 834–846.

Yang, Q. X., L. Wang, R. Yang, H. Stewenius and D. Nister (2009) Stereo matching with colour-weighted correlation, hierarchical belief propagation, and occlusion handling. *IEEE Transactions on Pattern Analysis & Machine Intelligence* **31**(3), 492–504.

Yang, Q., R. Yang, J. Davis & D. Nister (2007) Spatial-depth super resolution for range images. In: *2007 IEEE Conference on Computer Vision and Pattern Recognition*, pp. 1–8.

Yoon, K. J. and I. S. Kweon (2006) Adaptive support-weight approach for correspondence search. *IEEE Transactions on Pattern Analysis and Machine Intelligence* **28**(4), 650–656.

Yuan, X. (2008) On stereo model reconstitution in aerial photogrammetry. *Geo-spatial Information Science* **11**(4), 235–242.

Zabih, R. and J. Woodfill (2005) Non-parametric local transforms for computing visual correspondence. In: Eklundh, J. (eds) *Computer Vision — ECCV '94. ECCV 1994.* Lecture Notes in Computer Science, Vol. 801. Berlin, Heidelberg: Springer, 151–158.

Žbontar, J. and Y. LeCun (2015) Computing the stereo matching cost with a convolutional neural network. In: IEEE Conference on Computer Vision and Pattern Recognition, Boston, pp. 1593–1599.

Žbontar, J. and Y. LeCun (2016) Stereo Matching by training a convolutional neural network to compare image patches. *Journal of Machine Learning Research*, **17**, pp. 1–32.

Zhang, C., Z. Li, Y. Cheng and R. Cai (2015) MeshStereo: A global stereo model with mesh alignment regularization for view interpolation. In: International Conference on Computer Vision, Santiago, pp. 2057–2065.

Zhang, L. (2005) Automatic digital surface model (DSM) generation from linear array images. PhD thesis, Institute of Geodesy and Photogrammetry, Zurich, Switzerland.

Zhang, Q., R. Qin, X. Huang, Y. Fang and L. Liu (2015) Classification of ultra-high resolution orthophotos combined with DSM using a dual morphological top hat profile. *Remote Sensing* **7**(12), 16422–16440.

Zhang, Y., Y. Zhang, D. Mo, Y. Zhang & X. Li (2017) Direct digital surface model generation by semi-global vertical line locus matching. *Remote Sensi*ng **9**(3), 1–20.

Zhang, Z. (2000) A flexible new technique for camera calibration. *IEEE Transactions on Pattern Analysis and Machine Intelligence* **22**(11), 1330–1334.

Zhou, X. Z. and P. Boulanger (2012) Radiometric invariant stereo matching based on relative gradients. In: IEEE International Conference on Image Processing, Lake Buena Vista, FL, pp. 2989–2992.

Zhou, Z., N. Sang and X. Hu (2014) Global brightness and local contrast adaptive enhancement for low illumination colour image. *Optik* **125**(6), 1795–1799.

Zhu, Q., Y. Zhang, B. Wu and Y. Zhang (2010) Multiple close-range image matching based on a self-adaptive triangle constraint. *Photogrammetric Record* **25**(132), 437–453.

Zhu, X. X. and R. Bamler (2010) Very high resolution spaceborne SAR tomography in urban environment. *IEEE Transactions on Geoscience and Remote Sensing* **48**(12), 4296–4308.

Zigelman, G., R. Kimmel and N. Kiryati (2002) Texture mapping using surface flattening via multidimensional scaling. *IEEE* Transactions on Visualization and Computer Graphics **8**(2), 198–207.

# Chapter 4
# Utilizing BIM as a resource for representation and management of indoor information

*Umit Isikdag and Sisi Zlatanova*

## 4.1 Introduction

Developments in information technologies have changed the way we manage the life cycle of buildings. Early approaches to automating the design phase involved the use of CAD tools to generate digital models of buildings. In parallel, GIS has been used to represent buildings within the macro scale of the urban environment. In recent years Building Information Modelling (BIM) technology has emerged as a solution to the poor interoperability of models and data standards in the construction industry. Today, this paradigm is accepted as a new method of information management and many projects in the US, Singapore, Dubai and the UK require BIM-based processes and the involvement of BIM managers (a profession that has emerged in last 10 years) in construction projects. A Building Information Model (BIM) can be defined as a digital model of a building which contains geometric and semantic information about all building elements. The model can be stored in a database or can be exchanged as a file. The model acts as a shared knowledge resource, which grows in parallel with the advancement of the project. Once construction is completed, the model continues to be available for use by the facility managers, who can use the model to manage the building. The philosophy of BIM advocates information sharing, collaboration and coordination based on a shared digital model. A BIM is a shared digital representation of a single building, founded on open standards and besides other purposes, intended for interoperability. Over the last ten years, a BIM standard (namely Industry Foundation Classes, IFC), which is defined by an international industrial alliance, has been evolving, supporting the various phases of the building's life cycle.

There are two main approaches to the generation of BIMs. The first one is the transitional approach, where a building model is created as a loosely coupled collection of drawings, each representing a portion of the complete BIM. These drawings are then aggregated through various mechanisms to generate additional views of the building, reports and schedules, as though there was a single BIM at the center. The second approach is the central project database approach, where the building model is stored in a central project database and managed using a software or integrated system. The strength of the second approach is the ability to organize every building element in one database, thus providing users with the opportunity to see immediately the results of any design revisions made to the model, have them reflected in the associated views, as well as to detect any coordination issues. The definitive characteristics of Building Information Models can be given as being (Isikdag, 2015):

- Object oriented: Most BIMs are defined in an object-oriented nature.
- Data-rich/Comprehensive: BIMs are data rich and comprehensive as they cover all physical and functional characteristics of the building.
- Three dimensional: BIMs always represent the geometry of the building in three dimensions.
- Spatially related: Spatial relationships between building elements are maintained in BIMs in a hierarchical manner (allowing for several representations, such as constructive solid geometry, sweeping and boundary representations).
- Rich in semantics: BIMs maintain a high level of semantic (functional) information about the building elements.
- View/Use-Form derivable: The views and Use-Forms are subsets or snapshots of the model that can be generated from the base information model. These can be automatically derived with respect to the user and application needs.

According to Przybyla (2010) BIM provides the following benefits to stakeholders of the construction process.

- Improved design process
- 3D visualization for the owner (static only)
- Coordination between disciplines
- Interference checking
- Facilitates energy efficiency and leadership in energy and environmental design (LEED)
- Automated quantity take-offs
- 4D scheduling
- Improved documentation of design intent
- Potentially used for fabrication of construction materials

Three approaches can be used to acquire (geometric) information about buildings and transfer it into the geospatial environment to help the development of 2D and 3D indoor models:

- The first approach is measuring and 3D reconstruction, where information about an existing building is collected from multiple sources, and indoor models are created with respect to an application. Drawbacks of 3D reconstruction are that modelling indoor spaces is a time-consuming process and, as the main purpose is acquiring the geometry of the building elements, the final model contains limited semantic information. Furthermore, it is often difficult – or even impossible – to collect information hidden in building elements.
- The second approach to the integration of buildings in the GI environment is accomplished through acquiring building information from 2D and 3D CAD drawings. The problems related to this approach are generally referred to as CAD-GIS integration problems. CAD systems are developed to model objects

that do not exist and they are designed to represent the maximum level of detail in terms of the geometry and the attributes of the model. On the other hand, GIS (geographical information systems) are developed to represent objects that already exist around us. Such data models represent the objects in a relatively abstract and simplified way (specifically in terms of geometry).

- The third approach for acquiring 3D building information is using digital BIMs and simplifying them (geometrically and semantically). As mentioned previously, BIMs are object-oriented, semantically rich, up-to-date and allow interrogation of the needed building parts in views. Similar to CAD models, BIMs are also focused on representing the maximum level of detail in terms of the geometry and attributes of the model. In recent years there have been various successful academic and industrial efforts to simplify BIMs and to implement them within the geospatial context (Cheng *et al.*, 2013, Lappiere and Cote, 2008). The algorithms for seamless conversion from BIMs into the geospatial environment are still in development.

As BIMs contain detailed geometric and semantic information about the building elements, indoor installations and furniture, the transfer of information from the BIMs to geospatial systems brings noticeable opportunities, which are listed in Section 4.2. The transfer of building information from BIMs to geospatial environments will also play a significant role in the automatic generation of indoor representations and indoor models. Some of the problems related to indoor modelling and mapping are summarized in Section 4.3, and these can be overcome by using BIMs and BIM Use-Forms, which are elaborated in Section 4.4. The impact of information transfer from the BIM and Use-Forms (as explained in Section 4.5) can be large: (i) towards realizing BIM opportunities, and (ii) also for providing solutions to indoor modelling and mapping problems.

## 4.2 Opportunities offered by the BIM

Isikdag and Zlatanova (2009) have identified opportunities for several knowledge domains that appear as a result of the transfer of information from the BIM to the 3D GIS environment. A selected set of opportunities from that study, which are related to indoor modelling and mapping, can be listed as follows:

### 4.2.1 Integration of logistics operations into large-scale construction process simulations

In the construction industry, 4D models are known as models that combine 3D models and time information on construction activities to demonstrate the progress of construction over time. 4D simulations are useful in understanding clashes in the process and they improve communication in the project management tasks. Construction enterprises might have different projects running in different parts of a city (or different cities): in this situation the enterprise needs to carry out logistics operations between its construction sites. Logistics operations are usually managed within a geospatial environment and, if required, the level and amount of geometric and semantic information can be transferred into the indoor models, the 4D simulations can be completed within a geospatial (geo-virtual) environment, and they can be extended to cover logistic operations.

### 4.2.2 Flood damage assessment and pre-renovation actions

Transfer of semantic information from BIMs into indoor models can help in assessing the damage caused by a flood. For example, questions such as, "Which elements of electrical wiring might be damaged?", "Which parts of the HVAC systems can be broken down?", "Which wall's covering needs to be replaced after the flood?", can be answered using the information acquired from the indoor model , i.e. without visiting the actual site. On the other hand, the assessment of damage after a disaster may well support the design stage of a renovation project (i.e. when the new owner of a building might ask to remove some building elements (i.e. walls, doors, windows) after assessing the post-disaster condition).

### 4.2.3 Facilitating evacuation activities

Emergency responders (e.g. fire fighters) are generally not aware of the interior structure, furniture and materials of a building. In many countries, the only information available within the fire brigade is a plan (indoor map) indicating the fire exits. Floor plans (if not damaged by the fire/flood) might be obtained from the various spots/or facility management office in that particular building, but they may be outdated and may not provide details on semantic information.

The implementation of BIMs within the geospatial context can provide emergency responders with tools that will help in three aspects:

1. These tools facilitate orienting (as the response personnel will know the geometry of the construction and possible exits in advance).
2. These tools can also enable safer indoor navigation and evacuation, as the emergency responders will be informed about the usage type of the different rooms – e.g. a room might contain flammable chemicals – and the materials of the building elements – e.g. a type of flooring that might get slippery when wet.
3. Such tools can support evacuation simulation applications, which will allow emergency responders to explore scenarios, train and prepare better for responding to emergencies (Xie *et al.*, 2022; Zhao *et al.*, 2022).

Such implementation can help in answering the question: "How many square metres of a building might be affected by an emergency in a given area?"

### 4.2.4 Indoor 3D geo-coding

Geo-coding is the process of assigning geographic identifiers (i.e. coordinates) to any type of information. The geo-coding process involves transforming descriptive location information into an absolute geographic reference. Today, the most common data that are geo-coded are postal addresses. The implementation of BIMs in a geospatial context will help in developing models and algorithms for indoor 3D geo-coding. In parallel, 3D geo-coding and address-matching, together with developments in indoor navigation, will facilitate all location-based services, including the delivery of goods and services (indoors).

### 4.2.5 Property valuation and tax evaluation

In some countries the property valuation and tax evaluation process requires geometric and semantic information on building elements/parts and furniture, such as the precise dimensions of rooms and the number and type of fixtures located within the house. In addition, any structural changes in the house or property can affect the tax valuation. The transfer of information from BIM to geoinformation models can facilitate the tax evaluation process at the urban level by providing up-to-date information on the current state of a building (in terms of the geometry of the building and other installations and movable objects) when required by the tax authority. Investigations with 3D GIS models have already shown great improvements in accuracy (Boeters *et al.*, 2015)

## 4.3 Problems in indoor modelling and mapping

Zlatanova *et al.* (2013) listed a series of problems related to indoor modelling and mapping. The problems are grouped under five domains:

- Acquisition and sensors. Collecting information for indoor environments has largely been adapted from acquisition methods used for outdoor environments. The most common indoor approach is to mount sensors or systems of sensors on robots, trolleys or backpacks, or even hand-held devices (Tango, ZEB). Many successful systems have been realized. The greatest challenge remains the localization and registration of acquired data (Kolodziej and Hjelm, 2006). Some success has been achieved by using simultaneous localization and mapping (SLAM), inertial measurement units (IMU), sonar sensors (Girard *et al.*, 2011) and wireless positioning systems, to name but a few technologies. As mentioned above, all these methods for data acquisition cannot record the internal invisible building components, such as columns, beams and utilities.
- Data structures and modelling. A large part of indoor modelling refers to the processing of raw data: editing, fusing, structuring and attributing. Editing involves the selection and cleaning of data. The various data sets are then fused to deliver a common scene of the indoor environment. In the structuring phase, features are extracted and organized in 3D object models, aiming to provide topology and semantics (Nikoohemat *et al.*, 2020). In each of these phases many challenging issues exit. Existing processing algorithms are still unable to provide automatic and robust algorithms that can deal with the complexity of the indoor environment.
- Visualization. With the help of computer graphics, significant progress has been realized in recent decades in the visualization of 3D models on desktop systems. However, with the increase in mobile devices, new visualization methods will have to be realized for devices with relatively small screens. More developments are expected also in the area of augmented reality (AR). Indoor environments are much more suitable for AR applications due to relatively constant light and the large number of features.
- Applications. The applications in indoor environments can be numerous: emergency response, facility management, shopping, tour guidance,

localization of assets and people, and so on. Almost every indoor application relies on indoor localization, tracking and navigation. The systems for indoor navigation are still very diverse and rely on predominantly 2D maps. New robust networks will need to be developed for automatic generation of user-adapted networks for path computation (Diakité and Zlatanova, 2016) or seamless indoor/outdoor way finding (Yan *et al.*, 2021). A large number of approaches have been developed as described in Zlatanova and Isikdag (2016) and many of them have been based on BIM (Liu *et al.*, 2021).

- Legal issues and standards. To be able to exchange and fuse data successfully, many standardization and legislation issues remain to be addressed. Despite progress, conversion between BIM and GIS models is still being researched and remains in the course of being developed. A number of other standards, such LADM and IndoorGML, need to be investigated and linked with BIM to support indoor applications (Zlatanova *et al.*, 2016; Alattas *et al.*, 2017).

The challenges to indoor modelling are summarized as in Table 4.1 (Zlatanova *et al.*, 2013).

As building information models contain valuable information regarding the geometry of the building, the building elements and the internal installations, information acquired from BIM can help in overcoming some of the problems of indoor modelling and the mapping domain. On the other hand, although BIM is viewed as a single shared repository of the building information which is stored in digital form, the implementation of a BIM approach in the business processes requires not only the utilization of a single shared repository but also the generation of application- and context-specific views of the building information model for each of the sub-processes or for each of the implementation focuses. These application-focused or implementation-focused specific views that are generated in real time can be termed the Use-Forms of BIM, where each Use-Form and BIM (i.e. the shared repository) is mutually dependent. The mutual dependency between the BIM and the Use-Form indicates that a BIM needs to be present in order for a Use-Form to be generated; on the other hand, a Use-Form of the BIM must be generated in order for a BIM to be utilized for a specific BIM use. Section 4.4 elaborates on the Use-Forms of BIM. Section 4.5 discusses how each Use-Form of the BIM can be beneficial in overcoming the problems related to indoor modelling and mapping.

## 4.4 Use-Forms of BIM

Olatunji (2012) has identified the different forms of project model and how they are put to use in the stages of the construction project life cycle. These are visualization; prefabrication and construction models; and models for design analysis. Others are procurement models, and facilities management and simulation models.

On the other hand, BIM uses are defined as "a method of applying building information modelling during a facility's lifecycle to achieve one or more specific objectives". The uses can be interpreted as use cases that cover multiple processes of BIM-based information management. As explained in PSU (2013), the defined

**Table 4.1** Current and emerging problems in indoor modelling and mapping.

| A. Acquisition and Sensors | B. Data Structures and Modelling | C. Visualization | D. Navigation | E. Applications | F. Legal Issues and Standards |
|---|---|---|---|---|---|
| Current Problem Group | | | | | |
| A1. Variable lighting conditions | B1. Software tools | C1. Web and mobile devices | D1. Navigation models | E1. Indoor modelling for crisis response | F1. Unification of outdoor and indoor models |
| A2. Variable occupancy, automated feature removal | B2. Diversity of indoor environments | C2. PoI and landmarks strategies | D2. Automated space subdivision | E2. Augmented systems | F2. The diversity of indoor environments |
| A3. Sensor fusion | | | D3. Optimal routing | E3. Gaming | |
| | | | | E4. Industrial applications | |
| Emerging Problem Group | | | | | |
| A4. Mobility | B3. Real-time modelling | C3. Real-time change visualization | D4. Navigation queries and multiplicity of targets | E5. Natural description of indoor environments (semantics) | F3. Security and levels of access |
| A5. Real-time acquisition of dynamic environments | B4. Dynamic abstraction | C4. Complexity visualization | D5. Travelling imperatives | E6. Real-time decision support | F4. Privacy |
| A6. Learning the composition of space | B5. Discovering the context of space | C5. Aural cues | D6. Discrete vs continuous navigation models | | F5. Copyright |
| | B6. Integration with GIS/BIM | | D7. Guidance | | |

BIM uses covered the overall life cycle of the project from inception to finalization. The project also elaborated on the concept of level of development (LOD), stating that, for each of the BIM uses, the level of development should be identified in order to maximize the benefit from the BIM use. The LOD describes the level of detail/granularity to which a model element is developed.

Isikdag (2015) mentioned that in order to support several phases and the stakeholders of the construction life cycle, several views of the BIM need to be generated. These views can be generated from files or databases by using application, database and web interfaces. These views can either be transient or persistent, depending on the need. The views are generated by the declaration of a model that is a subset of another model, or by declaration of a model that is derivable from another model. The original model is called the base model and the new model is called the view. The entities of the view are populated from the base model.

The persistent model views are generated by model translation and, under the following conditions, the (translated) model can be called the model view:

1. The view should not be a superset of a predefined (base) information model. The view can be a subset of the model or the model itself.
2. The view should provide a snapshot of the information model (or its subset).

If the model view is persistent then it will be stored in a physical file or a database; otherwise, if it is transient, the physical storage of the view is not necessary. Persistent model views can be used whenever there is a need to exchange a subset of BIM between various different domains or when there is a need to exchange a snapshot of the BIM in one stage of the project.

Implementation of the BIM uses in the project requires several views of BIM to be generated/derived to support specific BIM uses. If these BIM views are persistent, they can also be defined as the Use-Forms of BIM. The Use-Forms will provide a model at a LOD that is compliant with the implementation purpose of the Use-Form and the stage of the construction. For example, if a Use-Form X is generated for the concept design phase it can be in LOD1, while the same Use-Form X generated for a post-design phase can be in LOD 3–LOD4. These Use-Forms can serve for the specific BIM use they are derived for but, on the other hand, each BIM Use-Form (which can also be interpreted as an independent model) will provide several different opportunities for the modelling and management of indoor information. The following sections will elaborate first on each Use-Form of BIM and then on their potential for indoor modelling and management.

### 4.4.1 Cost-Management Use-Form

The cost management process covers the estimation of the cost of the building. In addition, the process involves the tracking of project costs during construction and the management of the operation costs of the building. The cost estimation process starts at the planning stage/pre-design stages, where the possible cost of the building is estimated by preliminary and Order of Magnitude (OOM) estimates. Estimates at the detail design level are known as definitive estimates and are mainly based on results of detailed quantity take-offs. These take-offs can be completed based on detailed design models by using software applications. The organization that opens the tender, as well as the bidders, also make use of detailed design models to make definite estimates during the bidding process. The change orders in the construction phase also require the procurement team to make cost estimates during the construction phase. Cost management activities not only involve cost estimation processes: they also include cost management efforts, which are focused on tracking the project budget with the help of the earned value management (EVM) approach. The cost management process also continues in the operation phase, where as-built models can be used as tools to assist facility managers when a repair or an installation is required. By utilizing the Cost Management Use-Form of the BIM, quantity take-offs can be performed; also, the cost estimates can be produced based on the BIM. The Cost-Management Use-Form should contain detailed 3D geometric information about building elements along with material properties for each building element. In addition, the unit cost of each material needs to be assigned to the model as semantic information. The completeness level of the model ranges from design-complete to as-built.

### 4.4.2 Current Situation Modelling Use-Form

This process is concerned with the current situation of the construction site and its peripherals. The current situation model should always be up-to-date and accurate, reflecting the latest situation of a construction site (from inception to the Facilities Management (FM) stage of a building). Current situation modelling can be done in various stages of the building's life cycle. In the early stages of the design, current situation models help the designers to understand the construction site better. For example, at the inception stage the current situation model will not provide a representation of the building/facility but it will provide representation of every other object and landform at the site. Several tools – such as laser scanning, UAVs, aerial photographs, digital models (including geospatial models) – can be used for information acquisition about the current situation of the site. Once a streaming or periodic information acquisition is established, the current situation model will then be constantly updated in the construction stage by using the detailed design model, the nD model and by using the other information acquisition mediums mentioned. The Current Situation Modelling Use-Form will contain a conceptual geometry of the building/facility in the early stages of the construction where geometry about indoors is not represented. The Use-Form in the construction stage can contain the geometric representation of indoors in a point-cloud form, where the level of semantic information in the model is either zero or very low. The completeness level of the model ranges from zero to concept-complete.

### 4.4.3 Design Authoring Use-Form

Design authoring involves the design development for the building, including the development of architectural, structural and MEP (mechanical/electrical and plumbing) design models, and the fusion of (validated) structural engineering models and MEP models with the architectural model following an iterative approach. Design authoring starts with the schematic design model, which is prepared in the pre-design phase. The model is evaluated with the owner in terms of compliance with the user requirements. Once approved by the owner, the architect and engineering teams work collaboratively to develop the detailed design model. The detailed design model includes detailed architectural, structural and MEP models. In the detail design phase, the BIM will act as the main facilitator of the collaboration, while BIM-based tools will help in the design coordination. The output of the detailed design phase includes the blueprints and documentation, which can be used for construction purposes. At the start of the bidding phase the owner conducts an evaluation of the detail design and, if there are parts of the detailed design that do not comply with the user requirements, changes in the design will be necessary. These changes are conducted as part of the design authoring process. Furthermore, once the bidding process is complete, the contractor will make its first evaluation of the detailed design and can discuss with the owner any design changes caused by difficulties in the procurement processes or difficulties in fabrication of some elements. The Design Authoring Use-Form model contains detailed 3D geometric and semantic information, including the structural, mechanical (i.e. HVAC, plumbing) and electrical elements of the building. The completeness level of the model ranges from concept-complete to design-complete.

### 4.4.4 Energy and Sustainability Analysis Use-Form

Energy analysis involves the analysis for calculation/simulation of the energy demand and carbon footprint of the building in order to optimize the overall operation costs and help with environmental protection. The process also covers checking the compliance of the design with energy efficiency standards and a sustainability evaluation of the building. The process starts in the pre-design phase, with calculation/simulation of the energy demand and simulation of design alternatives to reduce the carbon footprint of the building. The pre-design phase processes make use of mass models and energy demand simulation software for these calculations and simulations. These analyses take the location/position of the building and seasonal sun-angle changes into account. In the detailed design phase, the models of the building, which include detailed geometries and material information, make advanced energy analyses and simulations possible. These analyses takes building element geometries, topography, positioning of each element, materials and opening dimensions into account when calculating the energy demand of the building. The detailed design phase can involve checks, calculations and documentation to confirm that the building's energy performance complies with standards such as LEED and Building Research Establishment Environmental Assessment Methodology (BREAM). When conformance to a level of standard is required, the bidding phase documentation will be prepared taking into account both these standards and the detailed design model. In the construction phase the final design model will be required to aid procurement efforts in order to acquire materials and components complying with targeted sustainability criteria. The Use-Form provides a detailed 3D geometric/semantic architectural model, as well as the 3D geometric representations of mechanical (i.e. HVAC, plumbing) elements along with their semantic information. The completeness level of the model ranges from concept-complete to design-complete.

### 4.4.5 Regulation Compliance Checking Use-Form

Regulation compliance checking involves checking the various models at different stages of the building's life cycle in order to ensure that the (architectural, structural, mechanical) design and design implementations on site conform to the rules and regulations. Rule checking in the planning phase includes checks to enable the start of the pre-design stage, and to ensure that the dimensional attributes of the land plot comply with cadastral registry entries and development plan registry entries. The zoning regulations are obtained at this phase from the municipality. Rule checking during the pre-design phase is a process that is conducted by the surveyor and the architect to ensure that the pre-design conforms to the zoning regulations. Code checking during the detailed design phase is conducted with a building control consultant and the municipality to ensure that (architectural, structural, MEP) design conforms to national standards (including earthquake-resistant design regulations, LEED-BREAM, fire and elevator regulations) and rules relating to zoning. A construction permit is granted when these checks are complete. Checks during the construction phase are to ensure that the actual implementation of the design conforms to the design rules, the design regulations and zoning regulations. Construction phase checks include the testing of physical elements – for instance, to ensure the quality of the concrete and reinforcement bars – whereby the results can be integrated into the as-built model. This phase also includes checks to ensure that the implementation complies with energy efficiency requirements (and standards), disaster prevention regulations, and that infrastructure elements are correctly integrated with building components. The Use-Form provides a detailed 3D geometric/semantic architectural model, as well as 3D geometric representations of mechanical (i.e. HVAC, plumbing) and electrical elements along with their semantic information. The completeness level of the model is design-complete.

### 4.4.6 Structural and Mechanical/Electrical/Plumbing (MEP) Design Use-Form

This process covers the structural and MEP-related analysis/simulation in order to test/validate and finalize the structural and MEP models of the building. It involves model compliance checking against structural and MEP design standards/regulations. The process begins with the acquisition of all the design models developed. The developed models are enhanced, tested and validated using structural and MEP analysis software, including Finite Element Method (FEM) packages, fluid dynamics applications and lighting design tools. The process is iterative and incremental, whereby analysis and simulation tools are used in order to obtain optimal results. The overall process can be termed model-based optimization of structural and MEP models. At this phase there is considerable exchange of information with the design authoring process, as changes in the structural and MEP models will require changes in the architectural model. Similarly, the structural and MEP modelling team will also need to access the most up-to-date architectural model, in order to use it in calculations and simulation. Successful information exchanges between these two processes is key to the success of the overall design. A change order at the construction phase can require the re-design of a building element. This design change will in turn need to be reflected in the structural and MEP models. Once the structural and MEP design has been changed, based on the change orders, this stage will require analysis of the new situation and validation of the changed or new components. The Use-Form provides a detailed 3D geometric representation of structural components (e.g. columns, beams, load-bearing walls) along with mechanical (i.e. HVAC, plumbing) and electrical elements and their semantic information. The completeness level of the model ranges from concept-complete to design-complete.

### 4.4.7 3D Coordination Use-Form

3D coordination involves the management of BIM-based (synchronous and asynchronous) design processes and construction processes in order to prevent conflicts between the participating actors and clashes between the (virtual-virtual/virtual-real) building elements. 3D coordination efforts start with enabling and facilitating coordination in (synchronous and asynchronous) design processes, where architects and engineers participate in the process from various locations. The preparation of quantity surveys and cost estimates, which will form the basis for bidding documents, will require coordination between the members of the bidding team and the design team. At the construction stage, 3D coordination is carried out by the project management team in order to identify potential clashes between the completed parts of the construction and components that are scheduled to be put in place (i.e. to facilitate model conformance and to prevent the manufacture of non-conforming parts) and to prevent the occupation of working spaces by building elements due to mis-scheduling of the tasks. The Use-Form contains the 3D geometric/semantic architectural model, and the 3D geometric representations of mechanical (i.e. HVAC, plumbing) and electrical elements along with their semantic information and can also include other installations and furniture. The completeness level of the model ranges from concept-complete to as-built.

### 4.4.8 Virtual Mock-up Use-Form

Virtual mock-ups are generated in the late pre-construction phases, using 3D design/BIM tools to facilitate the manufacture and assembly of complex components of the building. The mock-ups produced can be used in various analyses and in validation/verification of the components prior to the actual on/off site manufacturing process. It is usual to begin preparing virtual mock-ups at the detailed design phase with input from suppliers. Once developed, the mock-ups are then evaluated by the architect and engineers. Compliance of the real components with the project context and guidelines is validated with the help of the virtual mock-ups. The team that prepares the bidding documents and evaluates bids utilizes virtual mock-ups to prepare better bidding documentation, to make more accurate price estimates and to achieve better evaluation of bids. At the construction stage virtual mock-ups can be required and used by the project management team to plan the storage of the real components and to re-evaluate the applicability of the components based on the parts of the construction that are already built at that date. The virtual mock-ups can be used at the operation stage as components of the as-built model, i.e. to provide information mainly about mechanical, HVAC and electrical components. The Use-Form provides a very detailed 3D geometric/semantic architectural model but only for some specific building elements or installations. The completeness level of the model is design-complete.

### 4.4.9 nD Modelling and Simulation Use-Form

This process covers the development of models for simulations related to time/cost/ emergency response/... and the conducting of these simulations. The 3D model developed during the detailed design phase forms the basis for nD analysis and simulations related to scheduling, cost optimization and intelligent routing. During this phase, if the information that is required does not reside inside the model, it can be acquired from external resources such as schedules. nD simulations can help contractors make more accurate bid submissions (i.e. submissions with more accurate time plans and costing) in the bidding phase. In the construction phase nD simulations, together with constant observation of earned values, enable stakeholders to make more realistic predictions about possible finish dates and the project's final budget. The Use-Form provides a detailed 3D geometric/semantic architectural model, as well as 3D geometric representations of mechanical (i.e. HVAC, plumbing) and electrical elements along with their semantic information. The completeness level of the model is design-complete.

### 4.4.10 Quality Management Use-Form

This is a process by which the owner ensures that the final product is of the agreed quality (i.e. fulfils the user requirements) that is documented by the construction contract. The construction contract, together with the detailed (final) design, is an agreement between the owner and the contractor, which is signed to guarantee that the user requirements will be fulfilled (also taking rules and regulations into consideration). The detailed design is a key component of the contract. On the other hand, there can be some changes (i.e. change orders) during the construction phase and the final implementation is depicted in the as-built model. The owner checks the as-built model together with the actual implementation to ensure that the final manufactured components comply with its requirements. The as-built model plays a key role in this evaluation process as many details of the actual implementation will not be visible to the human eye once the construction is finalized. The Use-Form provides a detailed 3D geometric/semantic architectural model, as well as 3D geometric representations of mechanical (i.e. HVAC, plumbing) elements along with their semantic information. The completeness level of the model ranges from design-complete to as-built.

### 4.4.11 As-built Use-Form

The as-built model can be regarded as the final model of the project. This model can be used in FM operations and as the building manual in the use phase. The as-built model is developed based on information considering all the updates at the construction and operation stages. All disciplines that take part in the building's life cycle eventually contribute to this model in one of the stages. The construction phase design processes involve the model updates (representing the actual implementation of the building elements) in order to transform the design model to the as-built model. The final product of this phase is the as-build model at LOD-AB (LOD-As Built). The as-built model is developed with updates to the design model, which is represented by the BIM. The BIM needs to be revised and updated according to the changes that occur during the construction and post-construction stages. The as-built model will then be derived automatically from the BIM when the construction of the building is finalized. The as-built model represents the highest and most accurate LOD that can be considered LOD-AB. The difference between LOD-AB from other design phase LODs, such as LOD 400, is that the LOD-AB not only represents the highest granularity but also the most up-to-date (accurate) information about the building. The Use-Form provides a detailed 3D geometric/semantic architectural model, as well as 3D geometric representations of mechanical (i.e. HVAC, plumbing) elements along with their semantic information. The completeness level of the model is as-built.

### 4.4.12 Earned Value Management (EVM) Use-Form

Earned value management is a technique for measuring the project performance based on budgeted cost. Earned value analysis enables predictions to be made on the final cost of a project and the finish date during the construction phase. The earned value management process requires the existence of an up-to-date project schedule and an actual costing, which can be derived from the nD model that is kept up to date during the construction phase. The information that is derived from the model can be used for EVM analysis. The Earned Value Management Use-Form should contain detailed 3D geometric information about building elements along with the material properties for each building element. In addition, the unit cost of each material needs to be assigned to the model as semantic information. The completeness level of the model ranges from design-complete to as-built.

### 4.4.13 Facility Management Use-Form

This Use-Form is aimed at the set of processes where building operations and facility management (FM) activities are carried out based on a digital representation of the building. The set of processes begins with the handover of the building to the user; the BIM continues to be maintained during the operation phase in order to support the everyday building operation procedures. In addition, FM tasks makes use of the BIM as an information resource to get detailed semantic information about different building elements at a very low level (e.g. information about furniture can be acquired at this stage). The model should also provide information on temporary HVAC elements, new spaces that are generated during the use, the state of the mechanical and electrical components of the building; and, with the integration of information from IoT nodes to the BIM, the model should become capable of representing indoor conditions (such as indoor air quality) in real time. The LOD-FM (i.e. Facilities Management) represents the model that resides in the BIM databases (DBs) and gets updated after the changes (i.e. fixing or replacement) to some components. The FM model can always be derived from LOD-FM when required. The Use-Form provides a detailed 3D geometric/semantic architectural model, as well as 3D geometric representations of mechanical (i.e. HVAC, plumbing) elements along with their semantic information. The completeness level of the model is as-built.

### 4.4.14 Demolition Use-Form

This use-form model is devised for the process that covers BIM-related activities linked to the demolition of a building. BIMs can be used to plan and manage disassembly and recycling operations. The model level that would be used is LOD-FM. The information concerning building elements geometry, materials and assembly details can be use in planning disassembly operations and related logistics. The Use-Form provides a detailed 3D geometric/semantic architectural model, as well as 3D geometric representations of mechanical (i.e. HVAC, plumbing) elements along with their semantic information. The completeness level of the model is as-built.

Table 4.2 below summarizes the properties of the 14 Use-Forms. It provides the geometric/semantic properties of each Use-Form and the completeness levels of the

**Table 4.2** Properties of Use-Forms.

| ID | Use Form | Geometric/Semantic Properties | Level of BIM Completeness |
|---|---|---|---|
| 1 | Cost-Management | Building Elements → Detailed 3D Geom.+ Semantics<br>Structural Elements → Detailed 3D Geom.+ Semantics<br>HVAC/Electrical → Detailed 3D Geom.+ Semantics<br>Value Adding Information → Unit Cost of Materials | Design-complete to as-built. |
| 2 | Current Situation Modelling | Building Elements → Point Cloud 3D Geom.<br>Structural Elements → Usually not exists<br>HVAC/Electrical → Usually not exists<br>Value Adding Information → Site Facilities | Zero to concept-complete. |
| 3 | Design Authoring | Building Elements → Detailed 3D Geom.+ Semantics<br>Structural Elements → Detailed 3D Geom.+ Semantics<br>HVAC/Electrical → Detailed 3D Geom.+ Semantics<br>Value Adding Information → Externally referenced notes on Design Decisions | Concept-complete to design-complete. |
| 4 | Energy & Sustainability Analysis | Building Elements → Detailed 3D Geom.+ Semantics<br>HVAC/Electrical → Optional<br>Value Adding Information → Detailed semantics of materials | Concept-complete to design-complete. |
| 5 | Regulation Compliance Checking | Building Elements → Detailed 3D Geom.+ Semantics<br>Structural Elements → Detailed 3D Geom.+ Semantics<br>HVAC/Electrical → Detailed 3D Geom.+ Semantics<br>Value Adding Information → Externally referenced notes on compliance checking results | Design-complete. |
| 6 | Structural & Mechanical, Electrical, Plumbing (MEP) Design | Building Elements → Optional<br>Structural Elements → Detailed 3D Geom.+ Semantics<br>HVAC/Electrical → Detailed 3D Geom.+ Semantics<br>Value Adding Information → Externally referenced notes on Design Decisions | Concept-complete to design-complete. |

*(Continued)*

**Table 4.2** Properties of Use-Forms. (*Continued*)

| ID | Use Form | Geometric/Semantic Properties | Level of BIM Completeness |
|---|---|---|---|
| 7 | 3D Coordination | Building Elements → Detailed 3D Geom.+ Semantics<br>Structural Elements → Detailed 3D Geom.+ Semantics<br>HVAC/Electrical → Detailed 3D Geom.+ Semantics<br>Value Adding Information → Clash Detection Tests Results | Concept-complete to as-built. |
| 8 | Virtual Mock-up | Building Elements → Detailed 3D Geom.+ Semantics of Selected Elements<br>Structural Elements → Usually not exists<br>HVAC/Electrical → Optional<br>Value Adding Information → Externally referenced notes on Product Manufacturer Details | Design-complete. |
| 9 | nD Modelling and Simulation | Building Elements → Detailed 3D Geom.+ Semantics<br>Structural Elements → Optional<br>HVAC/Electrical → Detailed 3D Geom.+ Semantics<br>Value Adding Information → Project Schedule + Unit Cost of Materials | Design-complete. |
| 10 | Quality Management | Building Elements → Detailed 3D Geom.+ Semantics<br>Structural Elements → Detailed 3D Geom.+ Semantics<br>HVAC/Electrical → Detailed 3D Geom.+ Semantics<br>Value Adding Information → Externally referenced QA/QC Reports | Design-complete to as-built. |
| 11 | As-Built | Building Elements → Detailed 3D Geom.+ Semantics<br>Structural Elements → Detailed 3D Geom.+ Semantics<br>HVAC/Electrical → Detailed 3D Geom.+ Semantics<br>Value Adding Information → Externally referenced notes on changes between the design model and the actual constructed elements | As-built. |
| 12 | Earned Value Management | Building Elements → Detailed 3D Geom.+ Semantics<br>Structural Elements → Optional<br>HVAC/Electrical → Optional<br>Value Adding Information → Project Schedule + Unit Cost of Materials | Design-complete to as-built. |
| 13 | Facility Management | Building Elements → Detailed 3D Geom.+ Semantics<br>Structural Elements → Optional<br>HVAC/Electrical → Detailed 3D Geom.+ Semantics<br>Value Adding Information → Externally referenced user manuals of the products | As-built. |
| 14 | Demolition | Building Elements → Detailed 3D Geom.+ Semantics<br>Structural Elements → Detailed 3D Geom.+ Semantics<br>HVAC/Electrical → Detailed 3D Geom.+ Semantics<br>Value Adding Information → Externally referenced notes on demolition process details | As-built. |

BIM (i.e. the model). Section 4.5 provides an analysis on how the models of these use-forms can be utilized and how this utilization will benefit from the BIM opportunities as well as how the problems related to indoor modelling and mapping can be overcome by this utilization.

## 4.5 An analysis of the impact of BIM Use-Forms

### 4.5.1 BIM-related opportunities provided by Use-Forms

Integration of logistics operations into large-scale construction process simulations will help construction professionals benefit from better planning of logistics operations by utilizing GIS where the construction site consists of multiple buildings that cover a relatively large region (such as construction of a large factory, station or airport). As some of the logistics operations will be conducted at (newly built) indoor parts of the facility, the following Use-Forms can be beneficial in supporting these operations:

- Cost Management Use-Form.
- 3D Coordination Use-Form.
- Quality Management Use-Form.
- Earned Value Management Use-Form.

The transfer of information from the BIM to the GIS models (such as indoor models) can facilitate tasks in flood damage assessment and pre-renovation processes. Most of the damage assessment tasks and pre-renovation planning will take place indoors and will require information from indoor models. The following Use-Forms can be helpful in providing this information to indoor models.

- Quality Management Use-Form
- As-built Use-Form
- Earned Value Management Use-Form
- Facility Management Use-Form

Emergency response operations will benefit from the indoor models populated from BIM Use-forms, as these models will help orient the evacuation staff, provide better indoor navigation decisions based on semantics obtained from the models, increase safety at operations, and will help the operations to be performed with fewer staff and in less time, based on emerging navigation and egress management approaches (see Figure 4.1). The following Use-Forms will be helpful in providing related information to indoor models:

- Quality Management Use-Form.
- As-built Use-Form.
- Earned Value Management Use-Form.
- Facility Management Use-Form.

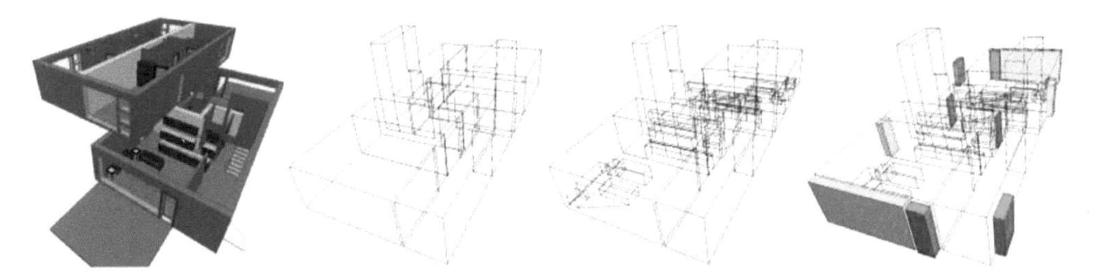

Figure 4.1 Extraction of the free space from a BIM model for the purpose of navigation. From left to right: original models; the space without furniture; free spaces considering the furniture; free spaces with doors (green) and windows (yellow) (Diakité and Zlatanova, 2016).

Indoor geo-coding will help in delivery operations in large facilities; the process will benefit from the indoor models populated from BIM Use-Forms. The following Use-Forms can aid in facilitating indoor geo-coding:

- Quality Management Use-Form.
- As-built Use-Form.
- Earned Value Management Use-Form.
- Facility Management Use-Form.

The valuation and property tax evaluation process in some countries will benefit from the geometric and semantic information contained in BIM and indoor models (see Figure 4.2). The following Use-Forms can aid in facilitating the property valuation related tasks:

- Cost Management Use-Form.
- Quality Management Use-Form.

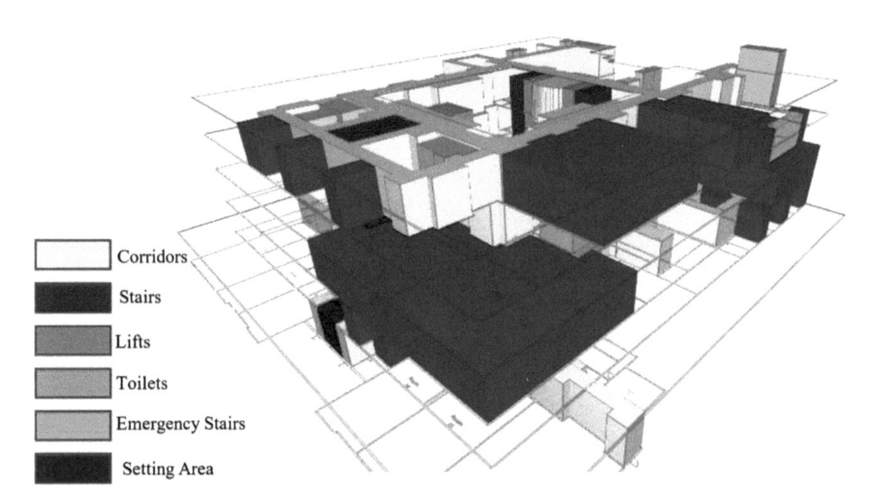


Figure 4.2 Indoor spaces modelled with respect to 'common' functional rights (Zlatanova *et al.*, 2016).

- As-built Use-Form.
- Earned Value Management Use-Form.
- Facility Management Use-Form.

In summary, indoor models derived from six different Use-Forms of BIM will provide opportunities for different tasks, ranging from indoor delivery operations to emergency response activities. The following section focuses on exploring how these Use-Forms can help in providing solutions related to problems in indoor modelling and mapping.

### 4.5.2 Overcoming indoor modelling and mapping problems utilizing BIM Use-Forms

Three main domains that can benefit from the implementation of Use-Forms to overcome problems related to indoor modelling and mapping are: data structures and modelling; applications; legal issues and standards. Several BIM Use-Forms can act as data sources of indoor models, where geometric and semantic information can be transferred. Readers are invited to revisit Table 4.1 prior to reading this section.

In general, problems related to (A) acquisitions and sensors cannot be overcome by using BIM Use-Forms; the Current Situation Modelling Use-Form can provide complementary information and contribute to the solution of (A3) the sensor fusion problem. As the Facility Management Use-Form will provide most up-to-date information about the current geometries and semantics of building installations it has the potential to contribute to the solution of (A5) real-time acquisition of dynamic environments problems. Information related to furniture in the 3D Coordination Use-Form and Facility Management Use Form will contribute to the problem of (A6) learning the composition of space.

BIM Use-Forms have the potential to help in the solution of problems in (B) data structures and modelling. The data structure of As-Built and Facility Management Use-Forms can help in the generation of taxonomies to overcome the (B2) the diversity of indoor environments problem. The real-time modelling problem can also be overcome similar to (A5), as the Facility Management Use-Form will provide most up-to-date information about the current geometries and semantics of building installations. These Use-Forms will also contribute to the solution of (B5) the discovering the context of space problem. The (B6) problem concerns the integration of BIM and GIS for facilitating the generation of indoor models. All 14 Use-Forms can contribute to BIM-GIS integration: depending on the purpose of the integration, each Use-Form can fulfil a different aspect of this integration. For instance, if the integration is done to facilitate indoor navigation of a facility in use, the As-Built and Facility Management Use-Forms would provide more appropriate information for the integration; on the other hand, if the integration is to be done to facilitate material delivery operations on the construction site, then nD Modelling and Simulation, 3D-Coordination, and Quality Management Use-Forms would provide more appropriate information for the integration.

The problems in domain (C), visualization, can benefit – to a limited level – from information acquisition from the BIM. For example, as a solution to problem (C2)

PoI and landmark strategies, the geometries and semantics of objects that would be visualized as landmarks can be acquired from the 3D Coordination Use-Form, and the As-Built and Facility Management Use-Forms. These landmarks can be furniture, signs, staircases, elevators and escalators, and building elements of entry and exit points. Use of the Facility Management Use-Form (as it will provide a real-time view of the building) will be very helpful in finding a solution to (C3) the real-time change visualization problem. BIM-based web service architectures can be used to serve and query the As-Built and Facility Management Use-Forms in order to help overcome (C4) the complexity visualization problem. The transient views of these Use-Forms or transient views of BIM tailored for specific visualization need may also be useful at this stage.

Problems related to (D), navigation, can be supported with the information derived from BIM Use-Forms. Demolition, As-Built and Facility Management Use-Forms can provide information for defining navigation models of indoor environments. Semantically enriched navigation models (such as network models) can be utilized and automatically generated by using these Use-Forms. These Use-Forms can also aid in (D2) automatic sub division of space, as BIM provides objects that model spaces as discrete (topologically closed) entities that are bound by surfaces. The Facility Management Use-Form will provide up-to-date information about the condition of each building element – for instance, if the elevators are in operation or not, or if certain rooms are accessible – and this information can be used to aid the solution of (D3) the optimal routing problem. If required by the domain, the As-Built and Facility Management Use-Forms will support the development of (D6) continuous navigation models. Applications related to guidance (D7), such as AR-VR applications, can make use of information derived also from these two Use-Forms.

The fifth domain of problems, (E), is related to applications that can benefit from indoor models. The 3D geometric and semantic information in the Facility Management Use Form and the As-Built Use-Form can act as information resources that can be utilized by indoor information models that support (E1) crisis response applications. In addition, augmented reality applications (E2) and gaming (E3) can benefit from the detailed representations of building elements and furniture that can be found in the 3D Coordination Use-Form and the Facility Management Use-Form. Depending on the domain in which they are implemented, the Quality Management Use-Form, the Earned Value Management Use-Form, the As-Built Use-Form, Facility Management Use-Form, Cost Management Use-Form, nD Modelling and Simulation Use-Form, and the Virtual Mock-up Use-Form can support different industrial applications (E4), ranging from indoor delivery operations to inventory/warehouse layout planning and production facility layout planning.

The final domain of problems is related to (F) legal issues and standards. The Current Situation Modelling Use-Form, the As-Built Use-Form, the Facility Management Use-Form and the 3D Coordination Use-Form will play a key role in the provision of 3D information about the surroundings of a building and the building model to city models based on BIM standard, which in turn would facilitate the unification of information in a standards-based manner.

## 4.6 Conclusions

In contrast with the previous coarse-grained view of understanding the BIM as a single digital 3D model representing an as-built situation, this study provides a more holistic view of BIM – as a shared data resource – and discusses how Use-Forms (i.e. sub-models) derived from BIM can be beneficial in overcoming the problems related to indoor modelling and mapping. Various opportunities are offered by the transfer of information from BIMs into 3D GIS environments in different fields, e.g. integration of logistics operations into large-scale construction process simulations, flood damage assessment and pre-renovation actions, facilitating evacuation activities, 3D-indoor geo-coding, property valuation and tax evaluation. Furthermore, many of the indoor-related modelling-related problems that were summarized in Zlatanova *et al.* (2013) can be facilitated by the transfer of information from BIM into 3D geoinformation models and also from the automatic generation of indoor models from BIM. The implementation of the BIM Uses in the project requires several views of BIM to be generated/derived to support specific BIM uses. If these BIM views are persistent, they can also be defined as the Use-Forms of BIM. Each BIM Use-Form (which can also be interpreted as an independent model) will provide several different opportunities for modelling and the management of indoor information. These opportunities have been elaborated throughout this chapter. It is important to note that BIMs will play a key role in the acquisition of semantic information for intelligent 3D city models.

## References


Alattas A, S. Zlatanova, P. van Oosterom P, E. Chatzinikolaou, C. Lemmen and K-J Li, 2017, Supporting Indoor Navigation Using Access Rights to Spaces Based on Combined Use of IndoorGML and LADM Models, *ISPRS International Journal of Geoinformation*, vol. 6 (12), pp. 384.

Boeters, R. K., Arroyo Ohori, F., Biljecki and S. Zlatanova, (2015) Automatically enhancing CityGML LOD2 models with a corresponding indoor geometry. *International Journal of Geographical Information Science*, **29**(12), December 2015, 2248–2268.

Cheng, J., Y. Deng and Q. Du (2013) Mapping between BIM models and 3D GIS city models of different level of detail. *Proceedings of the 13th International Conference on Construction Applications of Virtual Reality*, 30–31 October 2013, London, UK.

Diakité, A. A. and Zlatanova, S. (2016) Extraction of the 3D free space from building models for indoor navigation. *ISPRS Ann. Photogramm. Remote Sens. Spatial Inf. Sci.*, IV-2/W1, 241–248, doi:10.5194/isprs-annals-IV-2-W1-241-2016.

Girard, G., S. Côté, S. Zlatanova, Y. Barette, J. St-Pierre and P. van Oosterom (2011) Indoor pedestrian navigation using FootMounted IMU and portable ultrasound range sensors. *Sensors 2011*, Volume 11, 7606–7624.

Isikdag, U. (2015) *Enhanced Building Information Models: Using IoT Services and Integration Patterns*. Springer Briefs in Computer Science. Ham: Springer International Publishing.

Isikdag, U. and S. Zlatanova (2009) Towards defining a framework for automatic generation of buildings in CityGML using Building Information Models. In J. Lee and S. Zlatanova (eds) *3D Geo-information Sciences*, Lecture Notes in Geoinformation and Cartography (pp.79–96). Berlin & Heidelberg: Springer.

Kolodziej, K. W. and J. Hjelm (2006) *Local Positioning Systems: LBS Applications and Services*. Boca Raton, FL: CRC Press.

Kreider, R. G., & Messner, J. I. (2013). *The uses of BIM: Classifying and selecting BIM uses.* The Pennsylvania State University, 0–22.

Lappiere, A. and P. Cote (2008) Using open web services for urban data management: A testbed resulting from an OGC initiative for offering standard CAD/GIS/BIM services. In M. Rumor, V. Coors, E. M. Fendel and S. Zlatanova (eds) *Urban and Regional Data Management* (pp. 381–393). London: CRC Press.

Liu, L. B. Li, S. Zlatanova, P. van Oosterom, 2021, Indoor navigation supported by the Industry Foundation Classes (IFC): A survey, *Automation in Construction*, Vol 121, January 2021, 10436.

Nikoohemat, S., A. A. Diakité, S. Zlatanova and G. Vosselman, 2020. Indoor 3D reconstruction from point clouds for optimal routing in complex buildings to support disaster management, *Automation in Construction*, Volume 113, May 2020, 103109.

Olatunji, O. A. (2012) The impact of building information modelling on estimating practice: Analysis of perspectives from four organizational business models. PhD thesis, University of Newcastle, Newcastle, Australia.

Przybyla, J. (2010). The next frontier for BIM: interoperability with GIS. *Journal of Building Information Modelling*, (14–18). Available online at https://citeseerx.ist.psu.edu/document?repid=rep1&type=pdf&doi=316ec07020a65fa2d2b8aa64627d5de5029c3d6c.

PSU (2013) The Uses of BIM: Classifying and Selecting BIM Uses. Available online at https://scholar.google.com/citations?view_op=view_citation&hl=en&user=WZlM5QkAAAAJ&cstart=100&pagesize=100&sortby=pubdate&citation_for_view=WZlM5QkAAAAJ:JoZmwDi-zQgC.

Xie, R., S. Zlatanova and J. Lee, 2022, 3D indoor environments in pedestrian evacuation simulations, *Automation in Construction*, 144 (12), 104593.

Yan, J., S. Zlatanova, A. Diakité, 2021, A unified 3D space-based navigation model for seamless navigation in indoor and outdoor, *International Journal of Digital Earth*, 14(8), 985–1003.

Zhao, J., Q. Xu, S. Zlatanova, L. Liu, C. Ye and T. Feng, 2022, Weighted octree-based 3D indoor pathfinding for multiple locomotion types, *International Journal of Applied Earth Observation and Geoinformation*, 112 (8), 102900.

Zlatanova, S. and Isikdag, U. (2016) 3D Indoor models and their applications. In S. Shekhar, H. Xiong and X. Zhou (eds) *Encyclopedia of GIS* (pp. 1–12). Cham: Springer International.

Zlatanova, S., K-J. Li, C. Lemmens, and P. van Oosterom (2016) Indoor abstract spaces: Linking indoor GML and LADM. 5th International FIG 3D Cadastre Workshop, 18–20 October 2016, Athens, Greece, 317–328.

Zlatanova, S., G. Sithole, M. Nakagawa, and Q. Zhu (2013) Problems in indoor mapping and modelling. *The International Archives of the Photogrammetry, Remote Sensing and Spatial Information Sciences*, Volume XL-4/W4, 2013. 51– 55 December 2013, Cape Town, South Africa.

Chapter 5

# A review of existing tools and methods for the management and visualization of 3D city models

*Mehmet Buyukdemircioglu and Sultan Kocaman*

## 5.1 Introduction

3D city models have been used extensively by various disciplines and organizations in urban planning, disaster management, navigation, etc. and are still increasing steadily in newer fields like accurate assessment of the protected areas (Tezel *et al.*, 2019) or local 3D applications such as university campuses (Buyukdemircioglu and Kocaman, 2018a). Generating and visualizing 3D city models requires considerable manual labor, usage of different software, tools and knowledge. For instance, city models usually consist of buildings, but are generated and visualized together with a digital terrain model (DTM) and other city objects like bridges, roads, city furniture and similar urban features as these models describe the general shape and structure of the city. In addition, semantic information is more and more often included in order to perform analyses and queries at a higher level, thus requiring more storage (Ujang *et al.*, 2014) and effort (Kolbe, 2009) compared to traditional city models. Given the complexity of city models and the increasing user demand for high performance, flexibility and interoperability, the choice of the technology for their management and visualization is a crucial task.

In this chapter an overview of existing tools and methods for 3D city model data management and visualization is provided. After a brief recap on city model contents in Section 2, an overview of data storage and management formats is given in Chapter 3, and several popular software solutions are examined. Commonly used visualization engines and formats are described in Sections 4 and 5, respectively, and their advantages and limitations are discussed accordingly. The final section focuses on conversion tools for 3D city model storage and visualization formats.

## 5.2 Content of city models

3D city model generation and utilization is of great interest to municipalities, companies and researchers due to its wide usage areas and various applications. The main components of a 3D city model are usually digital terrain models (DTMs), building models, street-space models and green-space models (Döllner *et al.*, 2006), but they can also contain other city objects like bridges, trees and roads. On the other hand, most 3D city model applications are mainly focused on building models, and other contents of the 3D city models like roads, water bodies and vegetation are often ignored (Kumar *et al.*, 2018). 3D city models should be generated and visualized as a whole, together with other city model contents like city furniture, DTMs, basemaps, object attributes and textures for realism and attractiveness. In this section, the main components of the 3D city models are briefly summarized, and their role and importance are pointed out.

### 5.2.1 Buildings and city furniture

Buildings and city furniture are main components of a 3D city model. Building models can be generated from different data sources such as satellite images (Kocaman *et al.*, 2006), large-format aerial images (Buyukdemircioglu *et al.*, 2018), LiDAR point clouds (Kada and McKinley, 2009) and UAV images (Feifei *et al.*, 2012). With the recent developments in machine learning methods, 3D building reconstruction can be done using 3D point clouds and deep learning methods (Wichmann *et al.*, 2018). Buildings can be automatically textured from aerial or terrestrial images (Buyukdemircioglu *et al.*, 2018).

City furniture involves objects that increase the attractiveness and realistic look of a city model. These are immovable objects such as bus stops, traffic lights, lanterns, benches, advertising columns and delimitation stakes (OGC, 2010). Although city furniture is often used for visualization purposes, it can also be used for the determination of the optimal mass transportation or local structural conditions (Floros and Dimopoulou, 2016). The geometries of furniture models are in general designed in a CAD software or retrieved from existing furniture libraries.

City furniture objects are defined in standard CityGML schema as the thematic extension module "CityFurniture". They are usually instanced models of the same object and they can be modeled in different levels of detail (LoD) (0-4) and also have attributes like other objects of the city. Two scenes from a LoD3 model with various city furniture are provided in Figure 5.1.

### 5.2.2 Digital Terrain Model

The DTM is an important part of a 3D city model. Visualizing terrain together with the 3D city model provides a more realistic experience for users, increases the reliability

Figure 5.1 Traffic lights (left) and benches (right) from Gaziantep BizimSehir (Bizimsehir, 2019).

in decision making processes, and can be used for various applications like urban planning or disaster management.

A DTM can represent the terrain surface as regular grids or a TIN (Triangulated Irregular Network). However, grid models have several disadvantages compared to TINs (Kumar *et al.,* 2018). Regular grids require more storage space as the resolution of data increases compared to TINs (Stoter and Zlanatova, 2003), and they cannot represent complex surfaces with three dimensions as TINs. Terrain in CityGML schema is represented by "ReliefFeature" in different LoDs (0-4) (OGC, 2010) and can be stored as TIN. Kumar *et al.* (Kumar *et al.,* 2018) has developed a novel CityGML extension for the storage of TINs with large sizes in their study. They state that CityGML is not an efficient way for storing TINs, due to fact that the data size of the terrain can be very large for massive TINs, which is caused by data redundancy.

Terrain models can be visualized on the web together with buildings and other city objects. The resolutions of DTM and the LoD of buildings are not inter-dependent. Buildings with high LoD can be integrated with a low-resolution DTM, but for continuous representation of terrain, the "TerrainIntersectionCurve" (OGC, 2010) concept implemented in CityGML must be used to adjust the DTM. CesiumJS (Analytical Graphics Inc., 2018c), which is a widely-used open-source Javascript library, currently supports two formats for terrain visualization. Heightmap 1.0 format (Analytical Graphics Inc., 2015) represents the terrain as a regular grid, and Quantized-mesh 1.0 format (Analytical Graphics Inc., 2013) represents the terrain as TINs. Both formats use the multi-resolution LoD concept for efficient visualization. An example of visualizing a high-resolution terrain model and a textured 3D city model on the web is given in Figure 5.2.

Figure 5.2 1 m resolution DTM and textured 3D City model of Cesme, Turkey (Cesme3D, 2019).

### 5.2.3 Object semantics

3D city models are not only digital representations of the city geometries. A city model without semantics is just a visual model without any useful information about that city, which will allow only limited analyses and applications. On the other hand various applications (Biljecki *et al.*, 2015) require the model geometries and semantic information at the same time. Semantically enriched 3D city models can be used to gather, evaluate and store urban information (Billen *et al.*, 2017). Thus, the semantics can be used for 3D spatiotemporal queries and analyses. Integrating the models in a GIS (Geographical Information Systems) environment would increase the querying and simulation capabilities. Therefore, 3D city models should be enriched with semantic data, thereby transforming them into tools for understanding cities and increasing knowledge about them. Different formats exist for the storage of the geometric and semantic information, each one with advantages and disadvantages and a defined range of applicability.

CityGML is an interoperable semantic 3D city model exchange format (Gröger and Plümer, 2012), but it is not an appropriate format for visualization (Kolbe, 2009). CityGML needs to be converted into another format for high-performance visualization, but not all the visualization formats are capable of storing semantic information after conversion.

Google KML (Google Inc., 2004)/COLLADA format (Khronos Group, 2004), which is described in the following sections, can keep semantics to some extent. The attributes can be retrieved for each object from an external spreadsheet file, although it cannot meet the requirements for most web-based applications.

The Indexed 3D Scene Layers (I3S) (Esri, 2018) format can store attribute information, and can be visualized on commercial Esri ArcGIS or other supported web globes.

3D Tiles is a suitable solution for visualizing heterogeneous massive semantic 3D city models on the web (Murshed *et al.*, 2018). Different analyses, 3D queries and styling can be done on the web interface with the help of semantics and CesiumJS virtual globe. More detailed information about visualization formats is given in Section 5. An example of 3D query on CesiumJS can be seen in Figure 5.3.

Figure 5.3 Cesme: 3D city model coloured based on building area (Cesme3D, 2019).

# 5.3 Popular data storage and management software

## 5.3.1 Storage and management of 3D objects in DBMS

Modelling and mapping the third dimension can be done in all spatial databases (Döner and Bıyık, 2011). A spatial Database Management System (DBMS) allows users to store, manipulate, query and manage the geospatial data and also export to different format types. Storage and management of 3D geometries, attributes and textures in a DBMS is an important part of 3D city models. Spatial DBMS can handle huge amounts of data and allow multiple users to access the data at the same time. In this section, an overview of existing commercial and open-source spatial DBMS solutions for city model data storage and access is presented, and advantages and limitations are pointed out.

### 5.3.1.1 3DCityDB

3DCityDB (Yao *et al.*, 2018) is a software package for visualizing and managing CityGML data-based 3D city models. The software allows users to export CityGML data in KML/COLLADA formats for visualization. The data can be visualized in various applications such as CesiumJS and Google Earth. Setting up a 3DCityDB environment requires a running installation of several software tools, which are freely available. The workflow for visualizing city models with 3DCityDB software is shown in Figure 5.4.

In addition, the following software packages are specifically required in the installation environment:

- Java 8 Runtime Environment
- PostgreSQL and PostGIS
- 3DCityDB Importer/Exporter Tool, Scripts, and Web Map Client
- 3DCityDB Spreadsheet Generator Plugin
- Google Earth Pro
- Node.js

After the installation, the settings for PostgreSQL (PostgreSQL, 2019) and PostGIS (PostGIS, 2019) have to be made. In order to enable all spatial functions and data types, the PostGIS extension needs to be added to the new database by the database administrator (i.e. superuser). Then, the CityGML data schema is loaded to the created database instance.

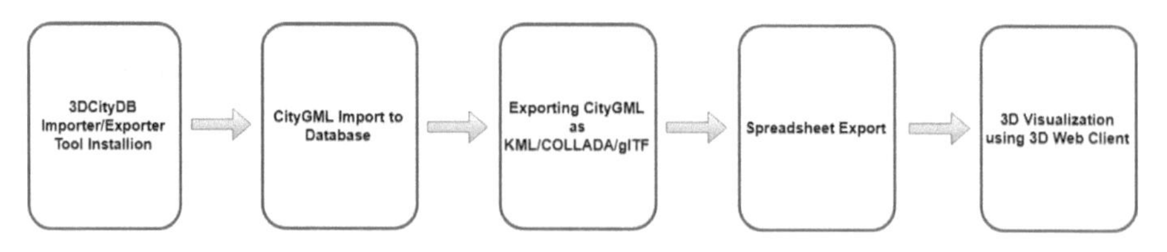


Figure 5.4 Workflow for visualizing with 3DCityDB software.

The 3DCityDB software package comes with a package of SQL (Structured Query Language) scripts to create the required schemas on the spatial database system (PostgreSQL/PostGIS) and with a group of scripts to manage the 3D city model stored in the database. 3DCityDB Importer/Exporter offers both a graphical user interface (GUI) and a command line interface (CLI). The database operations allow the user to manage and establish database connections and execute database operations.

Prior to the data import, the CityGML files should be validated against the CityGML XML schema. Per recommendations, CityGML files which have passed the validation process successfully shall be imported into the database. Otherwise, errors in the data may lead to unexpected behaviors. After the validation process is successfully completed, the CityGML file can be imported into the database. The spatial extent of the imported CityGML objects can also be easily determined by using the Importer/Exporter Tool.

The spatial data stored in the 3DCityDB can be exported in KML, COLLADA, and glTF formats for presentation, viewing, and visual inspection in different application areas such as CesiumJS Virtual Globe. Four different types of buildings exported as shown in Figure 5.5.

The 3DCityDB software package includes a Cesium-based web interface tool, named 3DCityDB-Web-Map-Client, for data visualization. This tool helps users to explore and visualize 3D city models with semantic data. Various changes and modifications have been made to the Cesium interface and source code to make it easier for users to visualize and explore 3D city models.

Figure 5.5 Different visualization forms of models from LoD0 to Textured LoD2.

Figure 5.6 Textured LoD2 buildings and attributes visualized on 3DCityDB-Web-Map-Client.

The 3DCityDB web client is a static web application written in HTML and JavaScript, and therefore can be easily deployed by uploading its files to a simple web server. The client comes with a lightweight JavaScript-based HTTP server that is mainly meant to test the functionality of the 3D web client on local machines. For running this web server, Node.js is required to be installed on the computer. After this last step, the client is ready for visualizing exported data on the web interface (see Figure 5.6).

#### 5.3.1.2 CityGRID

CityGRID (UVM Systems, 2019) is a software package solution from UVM Systems for managing, visualizing and modelling 3D objects with high-resolution façade textures. The software package contains six different modules for different management requirements of city models.

CityGRID Manager is a management tool for maintaining building geometries of cities of districts on an Oracle or SQL Server DBMS. The data are stored as vector models in a structure compatible with CityGML format, and also stored as 3D polyline data for easy editing of 3D geometries. Thus, the imported city model can be updated and improved with CAD or GIS data. The main features of this module are:

- Storing buildings as surface models compatible with CityGML 2.0 schema
- Automatic consistency check
- Storing building geometries as 3D polyline which allows editing geometries
- Object-relational database schema.

CityGRID Administrator is the module for managing the database that 3D city models are stored in and for performing CityGRID Manager functions. Basic administrative tasks of the 3D city model database related to optimization and clean-up are performed within this module. For example:

- 3D City model and DTM Import/Export of pre-defined areas or building ID
- Data export in CityGRID XML, CityGML, KMZ, DXF or VRML formats
- Management of the 3D city model in the spatial database (attributes, versions etc.)
- Automatic correction of overlapping building parts.

#### 5.3.1.3 GeoRocket Geodatabase

GeoRocket Geodatabase (GeoRocket, 2019a) is a Java-based high-performance open-source geospatial data storage tool developed by Fraunhofer Institute for Computer Graphics Research IGD (Fraunhofer IGD, 2019). It can store CityGML, GML and GeoJSON files in a database. GeoRocket has two editions: GeoRocket Pro and GeoRocket Standard. The standard edition is open-source and offers limited functions compared to the professional edition.

GeoRocket Geotoolbox is an on-premise toolbox for CityGML to 3D Tiles and CityGML to I3S conversion. This toolbox is included in the GeoRocket Pro version. A comparison of the featires of the standard version and the professional version can be found on the GeoRocket web page (GeoRocket, 2019b).

#### 5.3.1.4 Esri CityEngine

Esri CityEngine (Esri, 2013) is a stand-alone commercial software for procedurally generating 3D city models and urban content. CityEngine is compatible with ESRI file geodatabase (also geometries with textures) and ESRI shapefile format. CityEngine supports KML, COLLADA, FBX, Wavefront OBJ and DXF formats for 3D data exchange. DTMs and basemaps can be used together with the published 3D city model by connecting to ArcGIS Online.

Building-model generation starts with the geodatabase creation. Then 2D information is added to the generated model and building models are extruded to 3D with selected roof geometries; finally, façades are created and textured. City furniture can be added too. The generated 3D city model is shared on the web and updated in the geodatabase.

The 3D city model can be published on the web as a CityEngine web scene, which is a static form (cannot be edited online) of the generated model itself. All the generated building models, road networks and terrain are combined into a single CityEngine Web Scene (.3ws) file. The file is then uploaded to ArcGIS Online or portal, making it a browser-based 3D environment that allows user-oriented interaction. The published model can be viewed on the web by anyone without installing additional software. The model also can be shared as 360 VR for exploration with virtual reality glasses.

## 5.4 Visualization engines

3D City model visualization can be done using different engines and platforms for various purposes. Visualizing 3D city models on a web browser allows users to explore a model without installing any additional plugins or software. Different analyses can be done on web browsers with the help of semantic information. Visualizing on game engines will give the user more realistic and detailed experience compared to web browsers. These models can be explored with new technologies such as virtual reality. But with the newest developments, 3D city models can now be explored with mixed-reality technology using Microsoft Hololens. In this section, the widely used visualization engines are briefly described.

### 5.4.1 CesiumJS

Parallel to developments in web technologies and WebGL, 3D city model visualization on virtual globes became very popular and and has been extensively used in the last decade. CesiumJS (Analytical Graphics Inc., 2018c) is an open-source Javascript library for visualizing geospatial data on the web. It uses WebGL (Web Graphics Library) to increase the streaming performance of the datasets. It supports various types of geospatial data types, such as 3DTiles, terrain, imagery, point clouds and textured mesh. 3DTiles (Analytical Graphics Inc., 2019) is a data type developed by the Cesium team for visualizing 3D geospatial data with semantic information on the web. The 3DTiles format became an OGC standard in 2018 (OGC, 2019).

### 5.4.2 three.js

Three.js (Dirksen, 2013) is a Javascript library for visualizing 3D objects on a web browser. Just like CesiumJS, it works on web browsers without installing any plugins, and it uses WebGL technology for GPU acceleration. Although it supports more than 40 different file formats (three.js, 2019), some formats lack performance and are difficult to work with. The use of glTF (GL Transmission Format) is recommended since it is fully supported by three.js and efficient for visualization.

### 5.4.3 iTowns

iTowns (iTowns, 2016) is a three.js-based web framework for 3D geospatial data visualization. It supports various types of data such as 3DTiles, point clouds, GeoJSON, along with WMS/WMTS/TMS and elevation data.

### 5.4.4 GNOSIS

GNOSIS (GNOSIS, 2019) is a highly scalable object-oriented geospatial data visualization software development kit for 3D objects, imagery and terrain. It has been developed using eC (eC, 2019) programming language. It supports a variety of format types for terrain, imagery or vector data, such as GeoTIFF, ESRI Shapefile and ESRI ASCII Grid. 3DTiles and glTF file types are also supported for 3D objects.

### 5.4.5 ESRI ArcGIS API for JavaScript

ESRI ArcGIS API for JavaScript (Esri, 2016) is a commercial solution for publishing and visualizing 3D geospatial data on web. The published data are hosted on cloud, which is an advantage for fast sharing and updating of data, but is not suitable when working with sensitive data. API uses OGC standard ESRI I3S (Esri, 2018) as default file type. It is necessary to have an ESRI commercial license to use all functionalities of the API.

Figure 5.7 Coloured LiDAR point cloud of Bergama Town using PotreeJS library (Kalpakoglu *et al.*, 2018).

### 5.4.6 Potree

Potree (Schütz, 2016) is an open-source Javascript library for rendering large point clouds. It can handle point cloud data sets up to billions, generated from LiDAR or photogrammetric processing of optical images. It uses potree file format as default, which can be converted from other point cloud formats using an open-source tool potree converter (Schütz, 2016). Figure 5.7 shows part of a coloured LiDAR point cloud for Bergama Town, Turkey visualized with PotreeJS.

### 5.4.7 Google Earth

Google Earth (Google Inc., 2005) is a stand-alone software for rendering geospatial data on computers, smartphones or other portable devices such as tablets. 3D city models or landscapes along with GIS data or imagery can be loaded in same scene. It uses KML/COLLADA for visualizing 3D city models, which can be exported from CityGML using an open-source tool, 3DCityDB.

### 5.4.8 Unity Game Engine

Unity (Unity, 2019) is a cross-platform game engine for generating high-detailed models and exploring them in the generated scene. It supports virtual reality technology that gives users a more realistic experience. Projection systems, terrain model and imagery are supported as standard mapping functions.

### 5.4.9 Microsoft HoloLens

Mixed-reality hardware is a new technology solution for the urban planning experience. Microsoft HoloLens (Microsoft, 2019) is an untethered mixed reality hardware for visualizing and interacting with applications. GIS and Geoinformatics Lab group at ETH Zurich have developed a Microsoft HoloLens application called UrbanX (UrbanX, 2019), using City of Zurich open data for efficient and interactive urban planning.

## 5.5 3D Visualization formats

3D visualization has become a focus of interest for many industries and organizations in order to satisfy the requirements like administration, mobility and waste management of smart cities (Harbola and Coors, 2018). There are many ongoing studies of geospatial data visualization with semantic information and high visual detail on 3D virtual globes (Huang, 2017). 3D visualization has numerous advantages compared to 2D (Resch *et al.*, 2014). Since CityGML (OGC, 2010) is an exchange format for storing and exchanging semantic 3D city models, rather than a visualization format (Kolbe, 2009), it needs to be converted into other formats for high performance visualization on web globes or other platforms. The 3D city models can basically be visualized on web globes (Buyukdemircioglu and Kocaman, 2018), smartphones (Ellul and Altenbuchner, 2014) or game engines (Buyuksalih *et al.*, 2017). In this section, the most popular visualization formats are examined, and the advantages and limitations of each format is discussed.

### 5.5.1 3D Tiles

3D Tiles (Analytical Graphics Inc., 2019) is an OGC standard format for rendering and streaming large heterogeneous 2D/3D geospatial datasets, such as buildings, photogrammetric models, BIM/CAD models and point clouds, on CesiumJS virtual globe or other supported software packages. Loading large volumes of geospatial data or 3D city models on CesiumJS virtual globe as a single tile usually causes the web browsers to crash, therefore a tiling solution is needed. 3D Tiles, using Adaptive Quadtree Tiling, loads huge datasets as smaller parts and renders the datasets by dividing them into tiles, in an efficient and capable way. This approach reduces the amount of hardware resources used by the web browser and increases the streaming performance. 3D Tiles is based on geometric error for detail-level selection and an adjustable pixel defect, so that performance can be created for multiple zoom levels in the same view.

Murshed *et al.* (Murshed *et al.*, 2018) have identified some disadvantages of 3D Tiles as: no support for visualizing 4D city models; displaying the attribute window of the selected object requires manual coding in Javascript, which is an obstacle for dynamic customization and querying of the interface based on attributes of the dataset; and unavailability of open-source conversion tools. Another major issue is that Cesium currently does not support visualizing the data directly from a geodatabase so, if the dataset needs to be updated, a new dataset needs to be created or converted into 3D Tiles and replaced with the old one on the web server.

Streaming performance of the model varies on proper geometry and rendering optimizations, which reduces the size of the tileset. The gzipping support of 3D Tiles does not have any negative effect on the tileset, and increases rendering, streaming and

Figure 5.8 Textured CityGML of Cesme, Turkey, converted to 3D Tiles and visualized on Cesium (Buyukdemircioglu *et al.*, 2018b).

runtime performance of the 3D city model, while compressing texture coordinates, colours, vertex positions, normal and other generic attributes and reducing size. This also improves the efficiency and the speed of transmitting 3D content over the web. Open-source tool "3D Tiles Tools" (Analytical Graphics Inc., 2018d) can be used for gzipping, ungzipping and converting 3D Tiles sets.

An example of 3D Tiles representation of a textured LoD2 model on Cesium is shown in Figure 5.8. Based on experience from the Cesme 3D Modelling and Visualisation Project (Buyukdemircioglu *et al.*, 2018b), the advantages and disadvantages of storing data in Cesium 3D Tiles format can be listed as:

+ High performance while visualizing large and complex datasets
+ Point cloud and textured mesh support
+ No external database or software requirements for implementation
+ Building geometry attributes and textures stored in the same file
+ Gzip support for reducing file size without any loss of quality
+ Supports picking, styling, and querying on dataset
– Visualization only on Cesium Web Globe, iTowns and GNOSIS
– No possibility to change attribute information of the buildings online
– No open-source conversion tool from CityGML to 3D Tiles.

### 5.5.2 KML/COLLADA

KML (Keyhole Markup Language) (Google Inc., 2004) is an XML-based file format that allows users to visualize geographic data on CesiumJS or other supported platforms. COLLADA (Collaborative Design Activity) (Khronos Group, 2004) is also an XML-based file format that represents 3D geometries. After integration of COLLADA, KML can be used to visualize 3D models on virtual globes. Although conversion of the CityGML data to COLLADA format makes it thematically designable in an advanced 3D software environment, the main disadvantage of conversion into this format is that all semantic data that are stored in CityGML data will no longer be available after the conversion. In other words, all possible database queries and aggregations regarding CityGML semantics and also graphical data structure will no longer be available in the COLLADA format environment. 3DCityDB software allows users to visualize CityGML data with the help of KML/COLLADA integration.

The advantages of storing the data in KML/COLLADA format are:

- City model can be visualized both on Google Earth or CesiumJS Web Globe
- Attribute information can be changed directly from the spreadsheet
- User-friendly interface and easy use of 3DCityDB-Web-Map-Client.

Some disadvantages of storing the data in KML/COLLADA format are:

- No query or styling capabilities
- Building geometry, attribute spreadsheet and textures are generated separately
- A DBMS is required
- Complex process steps and too many software needs to be used
- No gzip support for reducing file sizes for better performance.

### 5.5.3 Indexed 3D Scene Layers (I3S)

Indexed 3D Scene Layers (Esri, 2018) is an OGC standard for scalable 3D scene content for visualization and distribution and open specification for storage and transmission of large, heterogeneous 3D geospatial datasets as a scene layer package (SLPK). I3S is cloud, web and mobile friendly platform based on JSON, REST and modern web standards which supports 3D geospatial content, various coordinate systems along with a rich set of layer types. It supports 3D objects, integrated meshes, point clouds and building scene layers. Geospatial data is organized by using a hierarchical, node-based spatial index structure.

The main features of I3S can be listed as: web friendly (JSON + Typed Arrays), mobile friendly (works well with varying bandwidth), extensible (supports different types of content), declarative (reduces required implicit knowledge), efficient (uses spatial indexing for quick delivery), scalable (provides LoD support), protected (ensures that content is protected) and open (full specification publicly accessible) file type. CityGML data can be converted into I3S with discrete file geodatabase features with FME (Safe, 2019a) software. I3S can be published by ArcGIS Pro and ArcGIS online, and can be visualized on ArcGIS API for Javascript, ESRI City Engine and CesiumJS virtual globe. A textured CityGML model of Berlin rendered in ESRI ArcGIS using I3S is shown in Figure 5.9.

Figure 5.9 Textured CityGML model of Berlin rendered in ESRI ArcGIS using I3S (OGC, 2018).

## 5.6 Spatial ETL tools widely used for conversions

Interoperability, format conversion and data integration is another challenging aspect of 3D city models. There are both open-source and commercial ETL (Extract, Transform and Load) tools for exporting and converting 3D city models into intended formats. In this section, the most popular software for format conversion and exportation is reviewed.

### 5.6.1 FME

Safe software FME (Feature Manipulation Engine) (Safe, 2019a) is a commercial ETL software for 2D/3D spatial data manipulation, extraction and format conversion that is mainly used in GIS (Geographical Information Systems) and CAD (Computer Aided Design) applications. FME supports more than 400 format types (Safe, 2019b) for read-write, and conversions between these format types can be automated by FME Server application by setting up calendar schedules or triggering events, such as starting a new process when new data is uploaded by a user.

FME comes with a Cesium 3D point cloud reader/writer and Cesium 3D tiles writer by default. Direct conversion from CityGML and various file types to 3D Tiles can be done with a single FME workbench, which is the visual workflow (diagram) design interface. Along with the converted files, a "tileset.json" file is generated with the dataset, which defines the tile structure of the 3D dataset. With FME 2019, the software is integrated with the Cesium ION (Analytical Graphics Inc., 2018a) platform, which makes for easier format conversion and automates workflows by building an FME workbench. Upon completion of the Cesium ION workbench, an email from

FME is sent to the user with the generated app URL, where the newly converted tileset is streamed in CesiumJS. Other popular 3D visualization formats like ESRI I3S and Google KML are also supported. These conversions can be done directly from a geo-database like Oracle Spatial or PostgreSQL with PostGIS.

### 5.6.2 Cesium ION

Cesium ION (Analytical Graphics Inc., 2018a) is a web platform for converting and hosting geospatial 3D datasets. Large amounts of data can be tiled and streamed on CesiumJS and can be shared with other users and platforms. 3D geospatial datasets like textured CityGML and KML/COLLADA are supported and tiled into 3DTiles for streaming on web. These tiled models can be clamped to the terrain with the terrain clamping option, which adjusts the height of the CityGML objects such as Building, CityFurniture, Track, Road, Railway, etc.

The coordinate reference system must be defined in CityGML, KML/COLLADA, raster imagery or terrain data for geo-referencing. Photogrammetric datasets or point clouds without any geolocation information can be accurately geolocated, scaled and rotated onto a specific terrain with 3D Tiles Location Editor. Datasets with additional sidecar files like CityGML with textures or a raster file with projection information (.prj file) can be uploaded together or as a single zip file to avoid file management problems in big datasets and reducing uploaded file size.

GPU-accelerated 3D analyses on datasets can be done with the commercial solution Cesium ION SDK (Analytical Graphics Inc., 2018b), which extends CesiumJS. Distance measurement, area calculation, visibility analysis and other features are available and ready to use with user interface widgets (Figure 5.10).

### 5.6.3 3DCityDatabase

3DCityDatabase comes with KML/COLLADA/glTF exporter tool by default. CityGML datasets stored in a spatial database can be directly exported as KML/COLLADA/glTF. Direct format conversion from CityGML to KML/COLLADA/glTF is not supported by the software. Detailed information about 3DCityDatabase Importer/Exporter tool and KML/COLLADA format is given in Sections 3.1.1 and 5.2.

Figure 5.10 Measurement (left) and area calculation (right) on Cesium ION SDK (Analytical Graphics Inc., 2018b).

### 5.6.4 Agisoft Metashape

Agisoft Metashape (Agisoft, 2019) is a stand-alone photogrammetry software for digital image processing and 3D geospatial data production, which can process up to 50,000 photos as a single cluster. Dense point clouds, raster digital surface models, textured meshes and high-resolution orthophotos can be generated for use in GIS or photogrammetric applications. The software cannot generate 3D objects directly but generated data can be used for generating 3D models. Metashape uses GPU along with CPU for increasing and reducing process time.

Metashape can directly export point clouds and textured meshes in popular visualization formats. Point clouds can be exported in Cesium 3D Tiles or potree format, and textured meshes can be exported as Cesium 3D Tiles, Google Earth, KMZ or COLLADA files and visualized on suitable visualization engines.

## 5.7 Conclusions

This chapter provides a review of existing tools and methods for 3D city model visualization and management environments. The main elements of 3D city models, such as buildings, furniture and DTM are described briefly and the main system components, such as storage, visualization and format conversions, are elaborated. City models have become an essential data element for 3D GIS and their production and presentation requires a number of different software programs, which currently can be used with a high level of expertise although no fully automated process is possible for the time being. Fully automatic generation of building models and other city objects is still an open topic for many researchers. In addition, integration of semantic information to high-performance visualization formats and the dynamic updating of object geometries seem to be the main issues for the realization of an optimally running 3D GIS environment. Although interoperability between software and formats is an expected issue, due to the use of multiple software to fulfill the whole process, efforts made especially by OGC facilitate the solutions to the problems.

Last but not least, the areas of application of 3D city models are increasing each day, and it is essential that new tools are developed and existing ones are improved, so that the management requirements of smart cities can be met and the livability of cities can be raised worldwide.

## References

Agisoft (2019) Agisoft Metashape. Available from: https://www.agisoft.com/.

Analytical Graphics Inc. (2013) Quantized-mesh-1.0 terrain format. Available from: https://github.com/AnalyticalGraphicsInc/quantized-mesh/.

Analytical Graphics Inc. (2015) Heightmap 1.0 terrain format. Available from: https://github.com/AnalyticalGraphicsInc/cesium/wiki/heightmap-1.0.

Analytical Graphics Inc. (2018a) Cesium ION. Available from: https://cesium.com/ion/.

Analytical Graphics Inc. (2018b) Cesium ION SDK. Available from: https://cesium.com/ion-sdk/.

Analytical Graphics Inc. (2018c) CesiumJS – Geospatial 3D mapping and virtual globe platform. Available from: https://cesiumjs.org/.

Analytical Graphics Inc. (2018d) 3D tiles tools. Available from: https://github.com/AnalyticalGraphicsInc/3d-tiles-tools.

Analytical Graphics Inc. (2019) 3D Tiles – Specification for streaming massive heterogeneous 3D geospatial datasets. Available from: https://github.com/AnalyticalGraphicsInc/3d-tiles.

Biljecki, F., Stoter, J., Ledoux, H., Zlatanova, S. & Çöltekin, A. (2015) Applications of 3D city models: State of the art review. *ISPRS International Journal of Geo-Information,* 4, 2842–2889.

Billen, R., Cutting-Decelle, A.-F., Métral, C., Falquet, G., Zlatanova, S. & Marina, O. (2017) Challenges of semantic 3D city models: A contribution of the COST Research Action TU0801. *International Journal of 3-D Information Modelling*, 68–76.

Bizimsehir (2019) Gaziantep BizimSehir. Available from: http://bizimsehir.org/model.html.

Buyukdemircioglu, M. & Kocaman, S. (2018a) A 3D Campus application based on city models and Webgl. *ISPRS – International Archives of the Photogrammetry, Remote Sensing and Spatial Information Sciences,* XLII-5, 161–165.

Buyukdemircioglu, M., Kocaman, S. & Isikdag, U. (2018b) Semi-automatic 3D city model generation from large-format aerial images. *ISPRS International Journal of Geo-Information,* 7.

Buyuksalih, I., Bayburt, S., Buyuksalih, G., Baskaraca, A. P., Karim, H. & Rahman, A. A. (2017) 3d modelling and visualization based on the Unity Game Engine – Advantages and challenges. *ISPRS Annals of Photogrammetry, Remote Sensing and Spatial Information Sciences,* IV-4/W4, 161–166.

Cesme3D (2019) Cesme 3D City Model. Available from: http://www.cesme3d.com/.

Dirksen, J. (2013) *Learning Three. js: the JavaScript 3D library for WebGL.*

Döllner, J., Kolbe, T. H., Liecke, F., Sgouros, T. & Teichmann, K. (2006) The virtual 3D City model of Berlin – Managing, integrating, and communicating complex urban information. In *Proceedings of the 25th Urban Data Management Symposium Udms, Aalborg, Denmark, 15–17 May 2006.*

Döner, F. & Bıyık, C. (2011) Modelling and mapping third dimension in a spatial database. *International Journal of Digital Earth,* 4, 505–520.

eC (2019) eC Programming language. Available from: http://ec-lang.org/.

Ellul, C. & Altenbuchner, J. (2014) Investigating approaches to improving rendering performance of 3D city models on mobile devices. *Geo-spatial Information Science,* 17, 73–84.

Esri (2013) CityEngine. Available from: https://www.esri.com/en-us/arcgis/products/esri-cityengine/.

Esri (2016) ESRI ArcGIS API for JavaScript Available from: https://developers.arcgis.com/javascript/.

Esri (2018) i3s. Available from: https://github.com/Esri/i3s-spec.

Feifei, X., Zongjian, L., Dezhu, G. & Hua, L. (2012) Study on construction of 3D building based on UAV images. *International Archives of the Photogrammetry, Remote Sensing and Spatial Information Sciences,* Volume XXXIX-B1, XXII ISPRS Congress, 25 August – 01 September 2012, Melbourne, Australia.

Floros, G. & Dimopoulou, E. (2016) Investigating the enrichment of a 3d city model with various Citygml modules. *ISPRS – International Archives of the Photogrammetry, Remote Sensing and Spatial Information Sciences,* XLII-2/W2, 3–9.

Fraunhofer IGD (2019) Fraunhofer Institute for Computer Graphics Research IGD Available from: https://www.igd.fraunhofer.de/.

GeoRocket (2019a) GeoRocket Geodatabase. Available from: https://georocket.io/.

GeoRocket (2019b) Products | GeoRocket. Available from: https://georocket.io/products/.

GNOSIS (2019) GNOSIS. Available from: http://ecere.ca/gnosis/.

Google Inc. (2004) KML – Keyhole Markup Language. Available from: https://www.opengeospatial.org/standards/kml.

Google Inc. (2005) Google Earth. Available from: https://www.google.com/earth/.

Gröger, G. & Plümer, L. (2012) CityGML – Interoperable semantic 3D city models. *ISPRS Journal of Photogrammetry and Remote Sensing,* 71, 12–33.

Harbola, S. & Coors, V. (2018) Geo-visualisation and visual analytics for smart cities: A survey. *ISPRS – International Archives of the Photogrammetry, Remote Sensing and Spatial Information Sciences,* XLII-4/W11, 11–18.

Huang, Y.-K. (2017) Within skyline query processing in dynamic road networks. *ISPRS International Journal of Geo-Information,* 6.

iTowns (2016) iTowns – a Three.js-based JS/WebGL framework for 3D geospatial data visualization. Available from: http://www.itowns-project.org/.

Kada, M. & McKinley, L. (2009) 3D Building reconstruction from lidar based on a cell decomposition approach. *Int. Arch. Photogramm. Remote Sens.,* 38, 47–52.

Kalpakoglu, E., Kilic, L., Ural, S., Anbaroglu, B. & Kocaman, S. (2018) Hava Lidar Nokta Bulutu Verilerinin Analizi Ve Web Ortamında Sunumu. *VII. Uzaktan Algılama Ve CBS Sempozyumu UZALCBS 2018.*

Khronos Group (2004) COLLADA – COLLAborative Design Activity. Available from: https://www.khronos.org/collada/.

Kocaman, S., Zhang, L., Gruen, A. & Poli, D. (2006) 3D city modelling from high-resolution satellite images. *In Proceedings of the ISPRS Workshop Topographic Mapping from Space (with Special Emphasis on Small Satellites), Ankara, Turkey, 14–16 February 2006.*

Kolbe, T.H. (2009) Representing and exchanging 3D city models with CityGML. In *3D Geo-Information Sciences*, J. Lee and S. Zlatanova (eds), Berlin Heidelberg: Springer, pp. 15–31.

Kumar, K., Ledoux, H. & Stoter, J. (2018) Compactly representing massive terrain models as TINs in CityGML. *Transactions in GIS,* 22, 1152–1178.

Microsoft (2019) Microsoft HoloLens – Mixed reality technology for business. Available from: https://www.microsoft.com/en-us/hololens.

Murshed, S., Al-Hyari, A., Wendel, J. & Ansart, L. (2018) Design and implementation of a 4D web application for analytical visualization of smart city applications. *ISPRS International Journal of Geo-Information,* 7.

OGC (2010) OGC City Geography Markup Language (CityGML) encoding standard. Available from: https://www.opengeospatial.org/standards/citygml.

OGC (2018) OGC Testbed-13: 3D Tiles and I3S interoperability and performance ER. Available from: http://docs.opengeospatial.org/per/17-046.html.

OGC (2019) Open Geospatial Consortium (OGC) – 3DTiles. Available from: https://www.opengeospatial.org/standards/3DTiles.

PostGIS (2019) PostGIS – Spatial and geographic objects for PostgreSQL. Available from: https://postgis.net/.

PostgreSQL (2019) PostgreSQL: The world's most advanced open source relational database. Available from: https://www.postgresql.org/.

Resch, B., Wohlfahrt, R. & Wosniok, C. (2014) Web-based 4D visualization of marine geo-data using WebGL. *Cartography and Geographic Information Science,* 41, 235–247.

Safe (2019a) Safe software FME. Available from: https://www.safe.com/.

Safe (2019b) All applications & formats supported by FME. Available from: https://www.safe.com/fme/formats-matrix/.

Schütz, M. (2016) Potree: Rendering large point clouds in web browsers. *Institute of Computer Graphics and Algorithms,* TU Wien.

Stoter, J. & Zlanatova, S. (2003) 3D GIS, where are we standing? In *Proceedings of the ISPRS Joint Workshop on Spatial, Temporal and Multi-dimensional Data Modelling and Analysis. Québec City*. Canada: ISPRS.

Tezel, D., Buyukdemircioglu, M. & Kocaman, S. (2019) Accurate assessment of protected area boundaries for land use planning using 3D GIS. *Geocarto International*, 1–12.

three.js (2019) three.js-supported file formats. Available from: https://github.com/mrdoob/three.js/tree/dev/examples/js/loaders.

Ujang, U., Anton, F., Azri, S., Rahman, A. A. & Mioc, D. (2014) 3D Hilbert space filling curves in 3D city modelling for faster spatial queries. *Int. J. 3D Inf. Model.*, 3, 1–18.

Unity (2019) Unity Game Engine. Available from: https://unity3d.com/.

UrbanX (2019) UrbanX – Urban planning in mixed reality. Available from: https://urbanxgis.wordpress.com/.

UVM Systems (2019) CityGRID. Available from: http://www.uvmsystems.com/index.php/en/software.

Wichmann, A., Agoub, A. & Kada, M. (2018) Roofn3d: Deep learning training data for 3d building reconstruction. *ISPRS – International Archives of the Photogrammetry, Remote Sensing and Spatial Information Sciences,* XLII-2, 1191–1198.

Yao, Z., Nagel, C., Kunde, F., Hudra, G., Willkomm, P., Donaubauer, A., Adolphi, T. & Kolbe, T. H. (2018) 3DCityDB – A 3D geodatabase solution for the management, analysis, and visualization of semantic 3D city models based on CityGML. *Open Geospatial Data, Software and Standards,* 3.

# Chapter 6
# Representation of 3D and 4D city models

*Ken Arroyo Ohori, Hugo Ledoux, Jantien Stoter, Anna Labetski, Stelios Vitalis and Kavisha Kumar*

## 6.1 Introduction

Recent advances in technologies to collect 3D elevation information, e.g. LiDAR and photogrammetry (Haala and Rothermel, 2012; Mallet and Bretar, 2009; Shahzad and Zhu, 2015), have made it relatively easy for practitioners in different fields to automatically reconstruct 3D city models. These models typically contain buildings (Alexander *et al.*, 2009; Rottensteiner, 2003), as well as other city objects such as roads, overpasses, bridges and trees (Elberink, 2010). Their availability and applications are steadily increasing in the fields of city planning and environmental simulations such urban noise mapping, flood simulations, and disaster management (Biljecki *et al.*, 2015). Furthermore, since elevation data can be acquired every few months/years at a relatively low cost, so-called 4D city models—i.e. 3D city models covering the same region at different periods in time—can also be reconstructed. Historical 4D city models for the cities of Duisburg, Solothurn, Prague, Toul and Hamburg have been constructed using close-range photogrammetry techniques and a variety of data sources such as maps, photographs, paintings and wooden models (Kersten *et al.*, 2012). These models not only offer a view of these cities in the past: they are also useful for understanding the changes happening in a city over time. As further explained in Section 2.5, the fourth dimension can also be, instead of time, the different levels of detail (LoD) of a 3D city model.

It should be noticed that since different 3D/4D city models are in practice generated independently using different reconstruction methods, software and sensor data, the resulting models often significantly differ in their geometry (e.g. a collection of surfaces versus a volumetric representation), appearance and semantics (often not present at all). In addition, every application requires its own specific semantic and geometric LoD of the 3D data. Also, as these models are stored using different formats (e.g. text, XML or binary formats), their underlying data models often also differ. Substantial differences in models can even happen to models that were originally identical, through updates or through conversions between different formats. All these differences have profound influences in practice, such as the applications for which a 3D or 4D model can be used, the processing that is necessary to use it, and the likely errors that will be present in the end result.

It is thus important to be aware of the way in which 3D and 4D city models are actually modelled. In this chapter we focus on a specific aspect of this: the main data models and formats used in practice to store and exchange city models. We focus primarily on open standards in 3D (Section 2) and their possible extension to 4D. Since interoperability between different 3D/4D data models is required to facilitate seamless exchange and use of city models, we also describe some of the interoperability issues between different standards and some possible solutions to convert between them (Section 3).

Next, we explain briefly some of the difficulties around extending reconstruction methods from 3D to 4D, such as how to deal with topology (Section 4). Even though 4D modelling is in its infancy, we also discuss the issues that 4D representations will face in this regard. In Section 5 how the 3D and 4D city models can be visualised in practice.

## 6.2 Data models for modelling cities

### 6.2.1 Standard 3D visualisation formats

The 3D visualisation formats encode the geometry and appearance of a 3D model, as well as some ancillary information that is useful within a 3D modelling program, such as the model's scene (position of light sources and cameras), and/or a set of animations done with it. These are used in several fields, mostly for visualisation purposes. However, while they can be used for 3D city modelling, they do not model two very important aspects of a city model: its semantic and topological details.

The main standard 3D visualisation formats are:

*VRML* (*Virtual Reality Modelling Language*)[1] is an open text-based format for modelling dynamic and interactive 3D scenes. It was accepted as the first web-based 3D standard by the web 3D consortium in 1995 and is still a format widely supported by popular tools such as FME and Sketchup. The format offers 3D geometries, several texturing mechanisms, animations and scripting. VRML scenes can be easily compressed using utilities such as gzip for its distribution over the internet.

*X3D* (*Extensible 3D*)[2] was developed in 2001 as the successor to VRML. It is essentially an XML-based encoding of VRML with added functionalities, such as geospatial positioning, shaders and the capability to store scene information. The rendering and texturing of 3D models are based on the functionalities provided by the low-level graphics engines such as OpenGL and DirectX. It also has support for classic VRML, binary and JSON encoding. It is well supported by many browser plug-ins, as well as data generation and conversion tools.

*OBJ* (*Wavefront Object*)[3] is one of the most popular text-based formats in the 3D graphics community. It has simple 3D geometries such as polygons and triangles for storing 3D models. The OBJ format can also encode colour and texture information, which is stored in a separate file with the extension .MTL (Material Template Library). It does not support animation nor the modelling of scenes.

*COLLADA* (*COLLAborative Design Activity*)[4] is an open XML-based format by Khronos Group for the representation and exchange of 3D assets. Its focus is primarily on the exchange of geometry data and 3D scenery. COLLADA supports triangular mesh geometry and has extensive shading and texturing options, animations, physics, and even multiple version representations of the same asset. It is commonly used with KML (Keyhole Markup Language) to render 3D city models in Google Earth.

*glTF* (*GL Transmission Format*)[5] is a JSON-based open 3D format by Khronos Group for the exchange of 3D models. It also has binary encoding for storing mesh geometry and animation data. It provides compact representation of geometries, small file sizes, and less processing time for unpacking and using the models.

Other 3D visualisation formats include PLY (Polygon File Format), OFF (Object File Format), STL (STereoLithography), I3S (Indexed 3D Scene Layer), etc.

### 6.2.2 CityGML and CityJSON

While the aforementioned standard 3D modelling formats can be used to store cities, it is often more useful to store a 3D city model in a specially structured format with semantic information stored in a standardised way. In this manner, a semantic 3D city model can be readily processed and visualised using standard tools. CityGML (OGC, 2012) is the main such standard. Its aim is to define the basic standard classes that can be used to describe the most common types of objects present in a 3D city model, their components, their attributes and the relationships between different objects. CityGML is based on a number of standards from the ISO191xx family, and it is used both as an information model (e.g. in the form of UML models of its classes) and a data format, which is an XML-based representation of its classes using some definitions from Geography Markup Language (GML) (OGC, 2007).

Although most CityGML examples and datasets available focus only on buildings, CityGML allows us to represent other feature classes such as: relief, roads and railways, vegetation, bridges, and city furniture. These can be supplemented with textures and/or colours to give a better impression of their appearance. Specific relationships between different objects can also be stored using CityGML: for example, that a building is decomposed into three building parts, or that a building has a both a carport and a balcony.

CityGML defines different standard levels of detail (LoDs) for all 3D objects; Figure 6.1 shows the 5 possible levels for buildings, which is the best known concept of CityGML. These provide the possibility to represent objects for different applications and purposes and can be useful for visualisation.

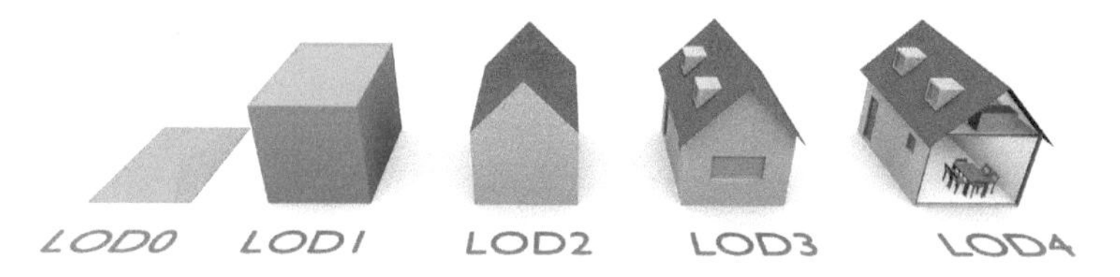


Figure 6.1 A building represented in LoD0 to LoD4 (figure from Biljecki *et al.*, 2014a). Notice there are four (volumetric) LoDs (LoD1 to LoD4); the non-volumetric LoD0 is a horizontal footprint and/or roof surface representation for buildings.

While CityGML prescribes a data model (feature classes and their attributes) for a 'generic' city (e.g. the function or year of construction of a building), it is possible to extend it for specific domains by defining ADEs (Application Domain Extensions). An ADE extends the CityGML schema with some new classes and/or new attributes for the existing features/elements. Examples of such extensions are: those for the energy demand of buildings (Agugiaro *et al.*, 2018) and for a country-specific data model for the Netherlands (van den Brink *et al.*, 2013). The main issue that ADEs are facing is that few software packages and libraries can read them automatically and thus process the information.

Apart from the most common XML-based CityGML, there are other formats that implement its data model. For instance, CityJSON[6] is a format that encodes a subset of the CityGML data model using JavaScript Object Notation (JSON). It offers an alternative to the GML encoding of CityGML, in which objects can be defined in a very large number of possible ways, and which can therefore be verbose and complex (and thus rather frustrating to work with). CityJSON aims at being easy to use, both for reading datasets and for creating them. It was designed with programmers in mind, so that tools and APIs supporting it can be quickly built. It is also designed to be compact, with a compression factor of around 7 when compared to XML-based CityGML, and it is also friendly for web and mobile development. A CityJSON object, representing a city, is as 'flat' as possible, i.e. the hierarchy of CityGML has been flattened out and only the city objects that are 'leaves' of this hierarchy are implemented. This considerably simplifies the storage of a city model compared to CityGML – and, furthermore, does not mean that information is lost. Another implementation is 3DcityDB,[7] which is an open-source database built upon Oracle Spatial[8] or PostGIS,[9] to store in a relational database the CityGML data model. It offers several extra functionalities to import/export city models from/to different formats, and it can thus help with interoperability.

### 6.2.3 InfraGML

The LandInfra conceptual model was developed by the OGC in cooperation with building SMART, with the aim of bridging the gap between two disciplines: architecture, engineering and construction (AEC), and the GIS industry (OGC, 2016c). It defines implementation-independent concepts for representing land and civil engineering infrastructure facilities such as buildings, roads, railways, and other features such as vegetation and terrain (OGC, 2016c).

Currently there is only one implementation of the LandInfra model: InfraGML, which is a GML-based implementation. It has 7 parts covering land features, facilities, alignment, roads, railways, surveys (including equipment, observations, and survey results), and land division (OGC, 2017). Each part of InfraGML is a separate OGC standard. *Core* is the mandatory part of the standard, which is extended by the other parts. It models the information contained in an InfraGML dataset, the definitions of the feature types, and the associations used in other parts.

InfraGML is expected to integrate easily with other OGC standards such as CityGML. There are significant overlaps between CityGML and InfraGML with respect to the modelling of spatial objects (OGC, 2016c). For instance, many CityGML objects are present as feature types in InfraGML such as Building (Building), Road (Road), ReliefFeature (LandSurface), WaterBody (LandFeature), VegetationObject (LandFeature), and LandUse (AdministrativeDivision) (OGC, 2016c). The alignment between InfraGML and CityGML standards is currently a topic of interest in the GIS community (Kumar, 2017).

### 6.2.4 BIM/IFC

The Industry Foundation Classes (IFC)[10] standard is an open data model used in the Building Information Modelling (BIM) domain for managing the information sharing processes throughout the life cycle of the building. While it was originally designed only for 3D buildings (and detailed representation of their interiors), the next version (IFC5) will support various infrastructure domains (e.g. roads, bridges, viaducts, etc.).

It has been adopted as the ISO 16739 international standard (ISO, 2013). Its geometric aspects are, however, mostly defined or derived from a different standard, ISO 10303 (ISO, 2014), which also specifies the STEP encoding that is most commonly used in IFC files.

IFC files can contain many types of classes (representing different aspects of a building, for instance a door, a staircase, a wall, the electric heater type, etc.), and the geometries used can be of different representation paradigms: (1) primitive instancing; (2) constructive solid geometry (CSG); (3) sweep volumes; (4) boundary-representation (b-rep). These four paradigms can be combined more or less freely; however, in practice, most IFC objects are built using sweep volumes and CSG (El-Mekawy and Őstman, 2010).

IFC models are highly relevant for 3D and 4D city modelling because, for many cities around the world, there is already a number of such models available, mostly for new buildings, manually made by architects. In practice, the automatic conversion between IFC models and other simple ones (e.g. CityGML) is a challenge (Arroyo Ohori *et al.*, 2017a; OGC, 2016b).

It should also be pointed out that the IFC standard is currently being extended so that other infrastructures of a city (e.g. roads and bridges) can be modelled.[11] Alignment between IFC and other relevant standards, e.g. CityGML, is also being investigated.

### 6.2.5 4D city modelling

#### 6.2.5.1 4D = 3D+time

The fourth dimension in 4D modelling often refers to the temporal aspect of the 3D data. Several researchers have tried to embed temporal aspects of geographic data in a 2D/3D+time data. The semantics of the time dimension included in these models vary from model to model but are often represented as a separate attribute either of an object or an event. An exception is the Space Time Cube (STC) (Kraak, 2008). However, the aim of the STC is merely to provide spatial insight into the temporal aspect rather than to realise a data structure to fully handle changes upon position, attributes and/or extent of the objects in a unified space-time-scale continuum. Arroyo Ohori *et al.* (2017c) describe spatio-temporal models that have been developed over the years.

The earliest approach to show time-related information of geo-data is as a series of separate snapshots where each snapshot is a representation of the dataset at one moment in time. Every object in the dataset is considered static until the time of the next snapshot. The main problem with this approach is that objects are represented multiple times, i.e. in all snapshots that they are part of, which causes data redundancy and possible data inconsistencies (due to edits in a snapshot that are not propagated into the others). Another problem is that when a change occurred is not recorded, i.e. it could have happened at any time between two snapshots. A solution to the problem of redundancy is to assign timestamps to every object, which demarcate the start and end of the period during which they exist. Therefore, objects only need to be represented again when they change. However, this approach also does not contain explicit events. In addition, in practice it appears ambiguous to decide when a change results in eliminating the object and creating a new one or keeping the existing one with changed attributes. Other approaches use events as principal entities, i.e. points in time where objects change, such as by keeping a list of changes per object. This makes it possible to know exactly when events occurred, and to identify and attach attributes to individual changes and events (e.g. what an event represents or why it occurred). All these solutions aim at capturing the change of objects (and their attributes).

Chaturvedi and Kolbe (2016) propose a new concept, called Dynamizer, allowing one to also model dynamic properties (i.e. temperature change over a day or year) of city objects and sensor observations. The Dynamizer concept can be applied to all city objects and is implemented as an Application Domain Extension for the CityGML standard. Dynamizers are intended to become part of the next version of CityGML (version 3.0).

#### 6.2.5.2 4D = 3D+LoD

As stated in the Introduction, 3D city models often differ in their geometry, appearance, LoD and semantics. This is due to the different reconstruction methods, software, and sensor data used to reconstruct the 3D city model. In addition, every application requires its own specific semantic and geometric LoD of the 3D data (Biljecki *et al.*, 2015). Take the urban object Building in Figure 6.1. Block models (LoD1) are sufficient for shadow simulations and the estimation of noise pollution, energy demand and fluid dynamics. Roof structures (LoD2) with information on the roof materials are needed for solar potential estimation, or in energy demand estimation. More detailed building models with information about windows and doors (LoD3) are important for estimating heat losses and for calculating the area available on vertical walls for solar panel installation. Building models that contain indoor spaces (LoD4) are required for indoor navigation and evacuation models.

LoD has a strong relationship to 'scale', as traditionally used for 2D maps, but in 3D it has a wider meaning (Biljecki *et al.*, 2014b). It not only reflects a ratio between measures in reality and on a map: it also reflects the amount of information (semantics and geometry) to be included to serve a specific application.

One of the strengths of CityGML is its support for five different LoDs. However, there are two major challenges of the CityGML LoDs. First, the standard allows for many alternatives for one LoD, and therefore in practice many variances of one LoD occur: for example, an LoD2 building with or without roof overhangs; with or without explicit modelling of roof, wall and floor; with or without dormers, etc. These variances are not further specified in CityGML but, as argued in Biljecki *et al.* (2016b), a more formal subdivision will significantly improve the quality of 3D city models. The authors therefore propose a further subdivision of the four CityGML LoD outdoor models for buildings (LoD0 to LoD3). The subdivision (resulting into 16 LoDs) is based on criteria such as the minimum size of an element (such as wall indentations and dormers) and the existence of specific elements such as roof overhangs, chimneys and openings (the latter two only for LoD3).

The second challenge of the CityGML LoDs is that while the LoDs for buildings are frequently studied and well known, the LoDs of other object types have received less attention. CityGML does not contain clear specifications for LoDs of other object types with the exception of tunnels and bridges, for which the LoDs are modelled similar to buildings. For vegetation, a SolitaryVegetationObject or a PlantCover may have a different geometry in each LoD, but the standard does not specify this further. Also LandUse objects (which can model the surface of the Earth as a subdivision into several polygons) may have different geometries in all LoD0–LoD4. But the CityGML standard does not provide any further guidance or specifications either, neither does it do so for the LoDs of the Relief class, which can be used to model the shape of the terrain.

Roads, another prominent object type in 3D city models, are modelled in CityGML under Transport and get a higher complexity at higher LoDs. But these are different from the LoDs of CityGML Buildings. At LoD0, they are represented as a linear network and, starting from LoD1, all transportation features are geometrically described by 3D surfaces, with a further thematic division of the road surface at higher LoDs. The CityGML standard has no information on the difference between the road surface representation at LoD1, LoD2, LoD3 or LoD4.

#### 6.2.5.3 Representation of 4D data models

So far there is no established data model for 3D+time and the 3D+LoD city models as described in the previous sections. They are therefore usually stored as separate 3D city models with some added metadata that explains the date of the model and possibly how the model was generated. While this is sufficient in some cases, it makes it difficult to automatically process such a model, requiring additional tasks such as performing a topological reconstruction of the model in 3D or 4D (Section 4).

Another challenge for 4D city models is their size—something that is easy to see if we compare the standard ways in which 2D and 3D city models are represented. In the Simple Features Specification (OGC, 2011), which reflects the prototypical way in which geometries are stored in 2D and 3D, independent 2D primitives are defined as sequences of points (with coordinates) that are connected using implicit line segments. When structures constructed from these 2D primitives are described (e.g. polyhedra), they are simply considered as sets of these 2D primitives.

Despite their apparent limitations, these types of structures can in fact be used to store objects of any dimension thanks to the Jordan-Brouwer theorem (Brouwer, 1911; Lebesgue, 1911). Just as independent 2D primitives are defined as sequences of points (with coordinates) that are connected using implicit line segments, 3D primitives can be represented as sets of such 2D elements, which are otherwise unlinked, and $n$D objects can be represented as aggregations of their $(n - 1)$ D-face bounding elements. Since every point in such an object can have any number of coordinates, an $n$D object can be embedded in $n$D space as well.

However, such representations become exponentially more inefficient as the dimension increases: lower-dimensional primitives need to be encoded multiple times, recursively once for each time they appear in a higher-dimensional primitive. This means that they are difficult to navigate and that even simple geometric and topological queries involve searching many objects (Hazelton, 1998). As an example, it is possible to consider Simple Features-like (OGC, 2011) representations of simple objects of various dimensions, as shown in Figure 6.2. A tesseract, or 4-cube, is the four-dimensional analogue of a square (in 2D) or cube (in 3D). It consists of 16 vertices, 32 edges, 24 faces, 8 volumes.

Considering how inefficient such representations can already be in 4D, it therefore makes more sense to use a topological representation for 4D city models (Arroyo Ohori, 2016). There are a number of ways in which this can be done in theory, such as with Nef polyhedra (Bieri and Nef, 1988), and with ordered topological models such as the cell-tuple (Brisson, 1993) and generalised/combinatorial maps (Lienhardt, 1994). These remain rather memory-intensive but they can still be more compact than a non-topological approach. However, research into applying these for 4D city models is still ongoing, and more work need to be done in order to make this a practical approach.

**(a) A square (2D), a cube (3D), and a tesseract (4D).**

```
[[0,0], [0,1], [1,1], [1,0], [0,0]]
```

**(b) square**

```
[[[0 0 0] [0 1 0] [1 1 0] [1 0 0] [0 0 0]]
 [[0 0 0] [0 1 0] [0 1 1] [0 0 1] [0 0 0]]
 [[0 1 0] [1 1 0] [1 1 1] [0 1 1] [0 1 0]]
 [[1 1 0] [1 0 0] [1 0 1] [1 1 1] [1 1 0]]
 [[1 0 0] [0 0 0] [0 0 1] [1 0 1] [1 0 0]]
 [[0 0 1] [0 1 1] [1 1 1] [1 0 1] [0 0 1]]]
```

*(Continued)*

**(c) cube**

```
[[[[0,0,0,0] , [0,1,0,0] , [1,1,0,0] , [1,0,0,0] , [0,0,0,0]],
   [[0,0,0,0] , [0,1,0,0] , [0,1,1,0] , [0,0,1,0] , [0,0,0,0]],
   [[0,1,0,0] , [1,1,0,0] , [1,1,1,0] , [0,1,1,0] , [0,1,0,0]],
   [[1,1,0,0] , [1,0,0,0] , [1,0,1,0] , [1,1,1,0] , [1,1,0,0]],
   [[1,0,0,0] , [0,0,0,0] , [0,0,1,0] , [1,0,1,0] , [1,0,0,0]],
   [[0,0,1,0] , [0,1,1,0] , [1,1,1,0] , [1,0,1,0] , [0,0,1,0]]],
  [[[0,0,0,0] , [0,1,0,0] , [1,1,0,0] , [1,0,0,0] , [0,0,0,0]],
   [[0,0,0,0] , [0,1,0,0] , [0,1,0,1] , [0,0,0,1] , [0,0,0,0]],
   [[0,1,0,0] , [1,1,0,0] , [1,1,0,1] , [0,1,0,1] , [0,1,0,0]],
   [[1,1,0,0] , [1,0,0,0] , [1,0,0,1] , [1,1,0,1] , [1,1,0,0]],
   [[1,0,0,0] , [0,0,0,0] , [0,0,0,1] , [1,0,0,1] , [1,0,0,0]],
   [[0,0,0,1] , [0,1,0,1] , [1,1,0,1] , [1,0,0,1] , [0,0,0,1]]],
  [[[0,0,0,0] , [0,1,0,0] , [0,1,1,0] , [0,0,1,0] , [0,0,0,0]],
   [[0,0,0,0] , [0,1,0,0] , [0,1,0,1] , [0,0,0,1] , [0,0,0,0]],
   [[0,1,0,0] , [0,1,1,0] , [0,1,1,1] , [0,1,0,1] , [0,1,0,0]],
   [[0,1,1,0] , [0,0,1,0] , [0,0,1,1] , [0,1,1,1] , [0,1,1,0]],
   [[0,0,1,0] , [0,0,0,0] , [0,0,0,1] , [0,0,1,1] , [0,0,1,0]],
   [[0,0,0,1] , [0,1,0,1] , [0,1,1,1] , [0,0,1,1] , [0,0,0,1]]],
  [[[0,1,0,0] , [1,1,0,0] , [1,1,1,0] , [0,1,1,0] , [0,1,0,0]],
   [[0,1,0,0] , [1,1,0,0] , [1,1,0,1] , [0,1,0,1] , [0,1,0,0]],
   [[1,1,0,0] , [1,1,1,0] , [1,1,1,1] , [1,1,0,1] , [1,1,0,0]],
   [[1,1,1,0] , [0,1,1,0] , [0,1,1,1] , [1,1,1,1] , [1,1,1,0]],
   [[0,1,1,0] , [0,1,0,0] , [0,1,0,1] , [0,1,1,1] , [0,1,1,0]],
   [[0,1,0,1] , [1,1,0,1] , [1,1,1,1] , [0,1,1,1] , [0,1,0,1]]],

  [[[1,1,0,0] , [1,0,0,0] , [1,0,1,0] , [1,1,1,0] , [1,1,0,0]],
   [[1,1,0,0] , [1,0,0,0] , [1,0,0,1] , [1,1,0,1] , [1,1,0,0]],
   [[1,0,0,0] , [1,0,1,0] , [1,0,1,1] , [1,0,0,1] , [1,0,0,0]],
   [[1,0,1,0] , [1,1,1,0] , [1,1,1,1] , [1,0,1,1] , [1,0,1,0]],
   [[1,1,1,0] , [1,1,0,0] , [1,1,0,1] , [1,1,1,1] , [1,1,1,0]],
   [[1,1,0,1] , [1,0,0,1] , [1,0,1,1] , [1,1,1,1] , [1,1,0,1]]],
  [[[1,0,0,0] , [0,0,0,0] , [0,0,1,0] , [1,0,1,0] , [1,0,0,0]],
   [[1,0,0,0] , [0,0,0,0] , [0,0,0,1] , [1,0,0,1] , [1,0,0,0]],
   [[0,0,0,0] , [0,0,1,0] , [0,0,1,1] , [0,0,0,1] , [0,0,0,0]],
   [[0,0,1,0] , [1,0,1,0] , [1,0,1,1] , [0,0,1,1] , [0,0,1,0]],
   [[1,0,1,0] , [1,0,0,0] , [1,0,0,1] , [1,0,1,1] , [1,0,1,0]],
   [[1,0,0,1] , [0,0,0,1] , [0,0,1,1] , [1,0,1,1] , [1,0,0,1]]],
  [[[0,0,1,0] , [0,1,1,0] , [1,1,1,0] , [1,0,1,0] , [0,0,1,0]],
   [[0,0,1,0] , [0,1,1,0] , [0,1,1,1] , [0,0,1,1] , [0,0,1,0]],
   [[0,1,1,0] , [1,1,1,0] , [1,1,1,1] , [0,1,1,1] , [0,1,1,0]],
   [[1,1,1,0] , [1,0,1,0] , [1,0,1,1] , [1,1,1,1] , [1,1,1,0]],
   [[1,0,1,0] , [0,0,1,0] , [0,0,1,1] , [1,0,1,1] , [1,0,1,0]],
```

Figure 6.2 A tesseract, or 4-cube, is the four-dimensional analogue of a square (in 2D) or cube (in 3D). It consists of 16 vertices, 32 edges, 24 faces, 8 volumes.

## 6.3 Interoperability between 3D city models

While there exist international standards providing unambiguous definitions of the 3D geometries in a 3D city model, as explained in Ledoux (2013, 2018), the majority of 3D GIS software ignores them and use its own definitions. The differences are fundamental: the abstract specification ISO 19107 (ISO, 2003) and its implementation in XML/GML (OGC, 2007) allow surfaces embedded in 3D space and solids to have inner boundaries (as is the case in 2D, where polygons can have 'holes'), while many software packages ignore these. Indeed, solids/volumes are often assumed to be bounded only by simple surfaces (thus forming a 2-manifold), while in fact they can be non-manifold objects. Such limitations can prevent practitioners from exchanging and converting datasets (since information is lost) and thus from using these in other software and applications.

Another issue that does not facilitate the exchange of 3D city models is that, in practice, their quality is often poor. As highlighted by Biljecki *et al.* (2016a), most openly available 3D city models contain geometric and topological errors, e.g. duplicate vertices, missing surfaces, self-intersecting volumes, etc. Often, these errors are not visible at the scale at which the datasets are visualised (Laurini and Milleret-Raffort, 1994) and, as a consequence, practitioners are not aware of the problem. But these errors prevent the datasets being used in other software and applications – see Nouvel *et al.* (2017), Steuer *et al.* (2015) and Bruse *et al.* (2015) for concrete examples in different application areas. While these geometric errors are many, they could be prevented if modelling software enforced the 3D geometries to be ISO 19107 compliant. Another solution to this problem is to use automatic repair algorithms; Attene *et al.* (2013) offer a survey of the methods to repair 3D models, as found in several disciplines. However, city models are not addressed, and the focus is on 'smooth surfaces', which seldom occur in a city context since buildings and bridges tend to have planar surfaces perpendicular to other surfaces.

Besides the geometry, the conversion of semantic 3D city models from one format to another is problematic because different data models might have incompatible semantics. One example is the differences between the object classes modelled in IFC and in CityGML. Take, for instance, a building (which both standards model): the mappings between the semantic classes are complex because different semantic information is attached to the geometrical primitives in the two models, and IFC has many more classes, whereas CityGML contains a limited number of classes structured in a hierarchy. A solution that has been proposed, among others by El-Mekawy *et al.* (2012), is to construct a common data model; this, however, does not allow one to use the files in already existing software, which would typically support either of these formats.

A vivid example of the current difficulties related to the interoperability of 3D city models is the automatic conversion between IFC models and CityGML models. While in theory this is attractive, since the 3D reconstruction phase could be skipped (since in many countries IFC files of new buildings are available), in practice the differences in semantics, coupled to the fact that different software and geometric modelling paradigms are used, have made the conversion impossible. OGC (2016a) and Arroyo Ohori *et al.* (2017a), among others, explain what the issues are that prevent one from automating the process, and provide recommendations so that both standards be better aligned.

It should be said that conversion between the purely geometrical formats (such as VRML, OBJ, glTF and OFF in Section 2.1) is usually not a major issue since these are often formed solely of triangles without any semantics, and the geometric errors are thus relatively easily solvable.

## 6.4 Construction of 3D and 4D models

### 6.4.1 Topological representations of 3D city models

One of the key benefits of 3D city models is that they can be processed automatically in a variety of applications (Biljecki *et al.*, 2015). This often involves the evaluation of topological relations between geometrical objects in 3D space. When these topological relationships are not known, we need to first identify the adjacency and intersection relationships between the objects (Guo *et al.*, 2010; Zlatanova, 2000).

As an alternative to expensive computations to obtain the topological relationships between objects on the fly, these can be precomputed and a topological data structure can be used in order to store those relationships directly in the dataset. By doing so, we can improve the calculation efficiency for many applications, such as doing network analysis of evacuation scenarios (Choi and Lee, 2009) or performing rendering on mobile devices (Ellul and Altenbuchner, 2014). The most commonly used topological data structures are the Doubly Connected Edge List (DCEL) or half-edge data structure (Muller and Preparata, 1978) and the quad-edge data structure (Guibas and Stolfi, 1985). Both have been used extensively in 2D GIS applications but their extension to 3D is not trivial.

A generalised model independent of dimension was originally proposed by Edmonds (1960), which later evolved into the data structure known as combinatorial maps (C-Maps) and was defined by Vince (1983). C-Maps by themselves only store abstract information regarding incidence and adjacency between the cells that it describes, but when they are combined with coordinate information on their vertices, they can represent any subdivision of 3D space with geometry. Those geometrically enhanced C-Maps are called linear cell complexes (LCCs). Feng *et al.* (2013) studied the concept of storing LCCs by proposing a solution for the compact storage of C-Maps for mesh data in 3D space. Damiand and Teillaud (2014) implemented a solution for storage and manipulation of dimension-independent data through LCCs.

LCCs have been used in the context of 3D city models and their applications. Horna *et al.* (2015) explore such an approach in modelling and simulation studies, where they describe the advantages of topological representations for 3D buildings. Diakité *et al.* (2015) have proposed EBM-LCC, a specific implementation of an LCC where attributes are used in order to describe 3D buildings that derive from BIM objects and LoD2 CityGML datasets. They also develop a method for creating EBM-LCCs from the topological reconstruction of buildings in order to automatically extract different LoDs from one main city object (Diakité *et al.*, 2014).

### 6.4.2 Creation of a topological 4D city model

Based on a series of 3D city models representing a city at different moments in time and which are all loaded into a 3D topological data structure, it is in theory possible to create a 4D city model that incorporates them all into one. However, it is worth noting that, in practice, the methods for doing so are still in development.

For instance, it is possible to take a 3D city model that exists at one point in time and then extrude it to 4D using the method described in Arroyo Ohori *et al.*

(2015b). This is the easiest method to load existing 2D or 3D data into a higher-dimensional structure, representing a set of cells that exist along a given dimension, such as a length of time or a range of scales. Unlike with other methods, it is also easy to guarantee that the output cell complex is valid and can be used as a base for further operations, such as dimension-independent generalisation algorithms. Once such an extruded model is created, it is possible to perform some operations in order to represent different situations, such as those expressed by transformations (i.e. translation, scaling and rotation) and the collapse of objects (Arroyo Ohori *et al.*, 2016).

Another possibility is based on the Jordan-Brouwer separation theorem (Brouwer, 1911; Lebesgue, 1911), since we know that a 4D object can be described based on a set of its bounding 3D objects. Since individual 3D objects are easier to describe than the 4D object, this can be used to subdivide a complex representation problem into a set of simpler, more intuitive ones (Arroyo Ohori *et al.*, 2014). Finally, one additional possibility is to link the series of 3D city models by identifying corresponding elements in different LoDs, deciding how these should be connected according to a linking scheme, and finally linking relevant 3-cells into 4-cells. Different linking schemes yield 4D models having different properties, such as objects that suddenly appear and disappear, gradually change in size, or morph into different objects along the fourth dimension (Arroyo Ohori *et al.*, 2015a).

### 6.4.3 Generalisation of 3D City Models

3D city models at lower levels of detail can be derived from models of a higher LoD: this is done through a process known as generalisation. Generalisation is motivated by the need to reduce the size and semantic complexity of a model to a level at which it can be utilised within a specific application while avoiding the loss of relevant information (Guercke *et al.*, 2009). Not all applications require the highest level of detail: rather, data needs are task-specific and data volume dependent (Baig and Rahman, 2012). Furthermore, there are often errors present in datasets that can occur due to different modelling software, different workflows used to produce the models and different approaches to quality assurance (Biljecki *et al.*, 2016a). Generalisation can therefore produce geometrically valid models that are tailored for usage within a specific application. Generalisation considers the geometry and/or semantics while targeting various city objects within a 3D city model. It is an iterative process and there is no one perfect output but, rather, various possibilities for a data user to experiment with.

Currently, due to the increasing availability of LoD2 models, there has been a higher focus on generalising from LoD2 to LoD1 (Figure 6.3). Generalisation of buildings from LoD2 to LoD1 can be accomplished by focusing on elements such as the floor plan and geometric reference. Simplifying the floor plan can be achieved by reducing the number of vertices through the decomposition of space along the major planes (Kada, 2008) or by adopting the Douglas-Peucker approach, which splits down line segments in a polygon until they are within a specific tolerance (Douglas and Peucker, 1973). The geometric reference of a building can be defined as the boundaries of the

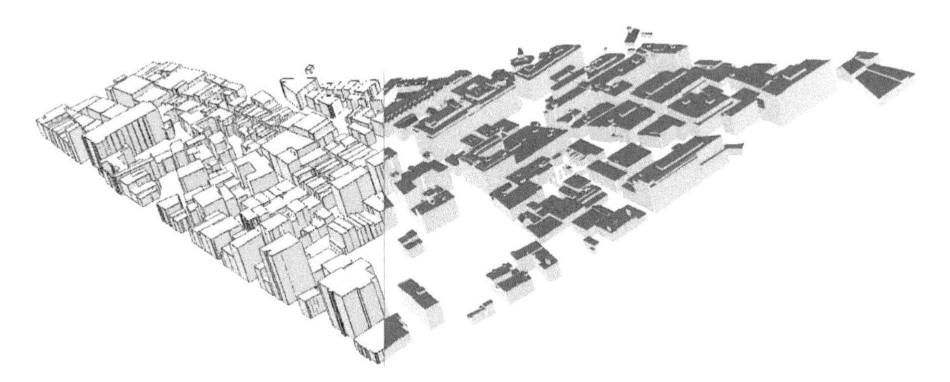

Figure 6.3 An example of generalisation from LoD2 (right) to LoD1 (left) on a subset of buildings in Montreal, Canada. Data courtesy of Portail Données Ouvertes, visualised with azul.

captured feature determined for a specific model (Biljecki *et al.*, 2016c). The vertical reference of a building at LoD1 can be set as either the minimum, maximum, mean, median, mode, percentage or percentile, based on the input roof values (Labetski *et al.*, 2017). Further, LoD1 models can either be generated by extruding from a generalised floor plan to its vertical reference, or the roof perimeter can instead be prioritised and 'downtrusion' can be employed to generate the building.

Harmonisation between city objects in the generalisation process is an important consideration because topological errors can occur if city objects are generalised without consideration of the surrounding objects. Errors can occur, such as neighbouring buildings with overlaps, loss of geographic relationships, and a misalignment with the underlying terrain. In CityGML the *terrain intersection curve* (TIC) denotes the exact position where the terrain touches a 3D object and is meant to ease the integration of city objects within the terrain (OGC, 2012).

## 6.5 Visualisation of 3D and 4D city models

### 6.5.1 3D

When 3D city models are stored in standard 3D modelling formats, they can be visualised in one of the many 3D viewers and 3D modellers that are readily available. While the exact features differ, there is a wide range of software available for this purpose. For instance, good choices include the mesh viewer and editor MeshLab[12] or the 3D modeller and renderer Blender,[13] both of which are free, open source and available in multiple platforms. In a web interface, these models can also be viewed with the help of libraries like three.js[14] and Cesium.[15]

When 3D city models are instead stored in CityGML or other semantic 3D city model formats, the options are more limited. However, such models can be viewed in Windows with the FZKViewer[16] or the Elyx 3D Viewer,[17] in Mac with azul[18] (Figure 6.4), or with cross-platform FME[19]. A list of software with CityGML support is available at https://www.citygml.org/software/.

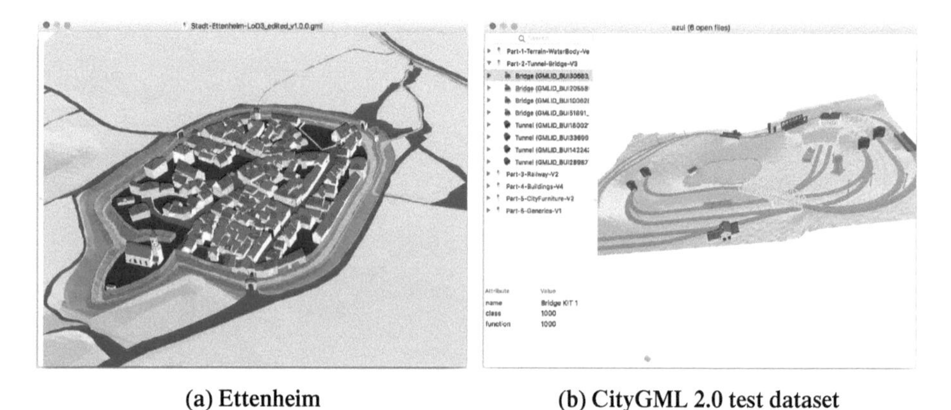

**(a) Ettenheim** **(b) CityGML 2.0 test dataset**

Figure 6.4 Two 3D city models visualised in azul. Different colours represent different semantics.

### 6.5.2 4D

Since 4D city models are still in their infancy, their visualisation is not a topic that has been studied much so far. Most visualisations are simply animations that show 3D city models evolving in time.

However, there is also a significant amount of work on the visualisation of general 4D objects, which can then equally serve to visualise 4D city models (Arroyo Ohori *et al.*, 2017b). This includes early work using visual metaphors of 4D space, such as *Flatland: A Romance of Many Dimensions* (Abbott, 1884) and *A New Era of Thought* (Hinton, 1888)

More recently, Beshers and Feiner (1988) describe a system that displays animating 4D objects that are rendered in real-time and uses colour intensity to provide a visual cue for the 4D depth. Banks (1992) describes a system that manipulates surfaces in 4D space, including interaction techniques and methods to deal with intersections, transparency and the silhouettes of every surface. Hanson and Cross (1993) describe a high-speed method to render surfaces in 4D space, with shading using a 4D light and occlusion, and Chu *et al.* (2009) also describe a system to visualise 2-manifolds and 3-manifolds embedded in 4D space and illuminated by 4D light sources. Notably, it uses a custom rendering pipeline that projects tetrahedra in 4D to volumetric images in 3D—analogous to how triangles in 3D are usually projected to 2D images. Arroyo Ohori *et al.* (2017b) show how a 4D model can be viewed in its entirety by projecting it to 3D using a variety of methods (Figure 6.5).

A different possible approach lies in using meaningful 3D cross-sections of a 4D dataset. For instance, Kageyama (2016) describes how to visualise 4D objects as a set of hyperplane slices. Bhaniramka *et al.* (2000) describe how to compute isosurfaces in dimensions higher than 3 using an algorithm similar to marching cubes. D'Zmura *et al.* (2000) describe a system that displays 3D cross-sections of a 4D virtual world one at a time.

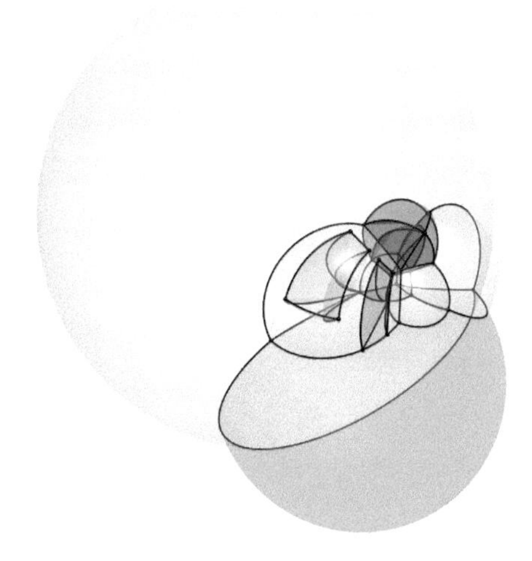

Figure 6.5 The model of a 4D house projected to 3D by first projecting inwards/outwards to the 3-sphere $S^3$, then stereographically to 3D.

## Notes

1 https://www.w3.org/MarkUp/VRML/
2 http://www.web3d.org/x3d/what-x3d
3 http://paulbourke.net/dataformats/obj/
4 https://www.khronos.org/collada/
5 https://www.khronos.org/gltf/
6 http://www.cityjson.org
7 https://www.3dcitydb.org
8 http://www.oracle.com/technetwork/database-options/spatialandgraph/overview/index.html
9 https://postgis.net/
10 http://www.buildingsmart-tech.org/specifications/ifc-releases
11 http://www.buildingsmart-tech.org/infrastructure
12 http://www.meshlab.net
13 https://www.blender.org
14 https://threejs.org
15 https://cesiumjs.org
16 https://www.iai.kit.edu/1302.php
17 https://1spatial.com/products/elyx/elyx-gis-platform/elyx-3d/
18 https://itunes.apple.com/nl/app/azul/id1173239678?mt=12
19 http://www.safe.com/

## References

Abbott, E. A. (1884) *Flatland: A Romance of Many Dimensions*. Seely & Co.

Agugiaro, G., Benner, J., Cipriano, P. and Nouvel, R. (2018) The energy application domain extension for CityGML: Enhancing interoperability for urban energy simulations. *Open Geospatial Data, Software and Standards*, 3(1).

Alexander, C., Smith-Voysey, S., Jarvis, C. and Tansey, K. (2009) Integrating building foot-prints and LiDAR elevation data to classify roof structures and visualise buildings. *Computers, Environment and Urban Systems*, 33(4): 285–292.

Arroyo Ohori, K. (2016) *Higher-dimensional modelling of geographic information.* PhD thesis, Delft University of Technology, April.

Arroyo Ohori, K., Damiand, G. and Ledoux, H. (2014) Constructing an n-dimensional cell complex from a soup of (n-1)-dimensional faces. In P. Gupta and C. Zaroliagis (eds), *Applied Algorithms. First International Conference, ICAA 2014, Kolkata, India, January 13-15, 2014. Proceedings*, volume 8321 of *Lecture Notes in Computer Science*, pages 37–48. Springer International Publishing Switzerland, Kolkata, India, January.

Arroyo Ohori, K., Ledoux, H., Biljecki, F. and Stoter, J. (2015a) Modelling a 3D city model and its levels of detail as a true 4D model. *ISPRS International Journal of Geo-Information*, (3): 1055–1075, July.

Arroyo Ohori, K., Ledoux, H. and Stoter, J. (2015b) A dimension-independent extrusion algorithm using generalised maps. *International Journal of Geographical Information Science*, 29:7, 1166–1186, DOI: 10.1080/13658816.2015.1010535.

Arroyo Ohori, K., Ledoux, H. and Stoter, J. (2016) Defining simple nD operations based on prismatic nD objects. In E. Dimopoulou and P. van Oosterom (eds), *11th 3D Geoinfo Conference*, vol. IV–2/W1 of *ISPRS Annals of the Photogrammetry, Remote Sensing and Spatial Information Sciences*, pp. 155–162, Athens, Greece, October 2016. ISPRS. ISSN: 2194–9042 (Print), 2194–9050 (Internet and USB).

Arroyo Ohori, K., Biljecki, F., Diakité, A., Krijnen, T., Ledoux, H. and Stoter, J. (2017a) Towards an integration of GIS and BIM data: What are the geometric and topological issues? In *ISPRS Annals of Photogrammetry, Remote Sensing and Spatial Information Sciences*, vol. IV-4/W5, pp. 1–8. doi: http://dx.doi.org/10.5194/ isprs-annals-IV-4-W5-1-2017.

Arroyo Ohori, K., Ledoux, H. and Stoter, J. (2017b) Visualising higher-dimensional space-time and space-scale objects as projections to R3. *PeerJ Computer Science*, July. doi: http://dx.doi.org/10.7717/peerj-cs.123. ISSN: 2376–5992.

Arroyo Ohori, K. Ledoux, H. and Stoter, J. (2017c) Modelling and manipulating spacetime objects in a true 4D model. *Journal of Spatial Information Science*, 14: 61–93.

Attene, M. Campen, M. and Kobbelt, L. (2013) Polygon mesh repairing: An application perspectice. *ACM Computing Surveys*, 45(2), Article 15.

Baig, S. U. and Rahman, A. A. (2012) Generalization and visualization of 3D building models in CityGML. *Lecture Notes in Geoinformation and Cartography*, p. 63–77, October 2012. Doi: 10.1007/978-3-642-29793-9 4. URL http://dx.doi.org/10.1007/ 978-3-642-29793-9_4.

Banks, D. (1992) Interactive manipulation and display of surfaces in four dimensions. In *I3D '92 Proceedings of the 1992 symposium on Interactive 3D graphics*, pp. 197–207, ACM.

Beshers, C. M. and Feiner, S. K. (1988) Real-time 4D animation on a 3D graphics workstation. In *Graphics Interface '88*, pp. 1–7. CHCCS/SCDHM.

Bhaniramka, P., Wenger, R. and Crawfis, R. (2000) Isosurfacing in higher dimensions. In *VIS '00 Proceedings of the conference on Visualization '00*. IEEE.

Bieri, H. and Nef, W. (1988) Elementary set operations with *d*-dimensional polyhedra. In H. Noltemeier (ed.), *Computational Geometry and its Applications*, vol. 333 of *Lecture Notes in Computer Science*, pp. 97–112. Berlin Heidelberg: Springer.

Biljecki, F., Ledoux, H. and Stoter, J (2014a) Redefining the level of detail for 3D models. *GIM International*, 28(11): 21–23. URL http://www.gim-international. Com/issues/articles/ id2167-Redefining_the_Level_of_Detail_for_D_Models. Html.

Biljecki, F., Ledoux, H., Stoter, J. and Zhao, J. (2014b) Formalisation of the level of detail in 3D city modelling. *Computers, Environment and Urban Systems*, 48: 1–15.

Biljecki, F., Stoter, J., Ledoux, H., Zlatanova, S. and Çöltekin, A. (2015) Applications of 3d city models: State of the art review. *ISPRS International Journal of Geo-Information*, 4(4): 2842–2889.

Biljecki, F., Ledoux, H., Du, X., Stoter, J., Soon, K. H. and Koon, V. H. S. (2016a) The most common geometric and semantic errors in citygml datasets, *ISPRS Ann. Photogramm. Remote Sens. Spatial Inf. Sci., IV-2/W1*, 13–22, https://doi.org/10.5194/isprs-annals-IV-2-W1-13-2016.

Biljecki, F., Ledoux, H. and Stoter, J. (2016b) An improved LOD specification for 3D building models. *Computers, Environment and Urban Systems*, 59: 25–37.

Biljecki, F., Ledoux, H., Stoter, J. and Vosselman, G. (2016c) The variants of an LOD of a 3D building model and their influence on spatial analyses. *ISPRS Journal of Photogrammetry and Remote Sensing*, 116: 42–54, June. doi: http://doi.org/10.1016/ j.isprsjprs.2016.03.003. URL http://dx.doi.org/10.1016/j.isprsjprs.2016.03. 003.

Brisson, E. (1993) Representing geometric structures in d dimensions: topology and order. *Discrete & Computational Geometry*, 9: 387–426.

Brouwer, L. E. J. (1911) Beweis des Jordanschen Satzes für den *n*-dimensionalen Raum. *Mathematische Annalen*, 71: 314–319.

Bruse, M., Nouvel, R., Wate, P., Kraut, V. and Coors, V. (2015) An energy-related CityGML ADE and its application for heating demand calculation. *International Journal of 3-D Information Modelling*, 4(3): 59–77.

Chaturvedi, K. and Kolbe, T. H. (2016) Integrating dynamic data and sensors with semantic 3D city models in the context of smart cities. *ISPRS Annals of Photogrammetry, Remote Sensing and Spatial Information Sciences*, IV-2/W1: 31–38.

Choi, J. and Lee, J. (2009) *3D Geo-Network for Agent-based Building Evacuation Simulation*, pp. 283–299. Berlin, Heidelberg: Springer. ISBN 978-3-540- 87395-2. Doi: 10.1007/978-3-540-87395-2 18.

Chu, A., Fu, C.-W., Hanson, A. J. and Heng, P.-A. (2009) GL4D: A GPU-based architecture for interactive 4D visualization. In *IEEE Transactions on Visualization and Computer Graphics*, vol. 15, pp. 1587–1594. IEEE.

Damiand, G. and Teillaud, M. (2014) A generic implementation of dD combinatorial maps in CGAL. *Procedia Engineering*, 82: 46–58. Doi: 10.1016/j.proeng.2014.10.372.

Diakité, A. A., Damiand, G. and Van Maercke, D. (2014) Topological reconstruction of complex 3D buildings and automatic extraction of levels of detail. In V. T. Gonzalo Besuievsky (ed.), *Eurographics Workshop on Urban Data Modelling and Visualisation*, pp. 25–30, Strasbourg, France, April 2014. Eurographics Association. Doi : 10.2312/ udmv.20141074. URL https://hal.archives-ouvertes.fr/hal-01011376.

Diakité, A. A., Damiand, G. and Gesquière, G. (2015) Automatic semantic labelling of 3d buildings based on geometric and topological information. *9th 3DGeoInfo Conference 2014 – Proceedings.*

Douglas, D. H. and Peucker, T. K. (1973) Algorithms for the reduction of the number of points required to represent a digitized line or its caricature. *Cartographica: The International Journal for Geographic Information and Geovisualization*, 10(2): 112–122, December. Doi: http://dx.doi.org/10.3138/FM57-6770-U75U-7727. URL https: //doi. org/10.3138%2Ffm57-6770-u75u-7727.

D'Zmura, M., Colantoni, P. and Seyranian, G. (2000) Virtual environments with four or more spatial dimensions. *Presence*, 9(6): 616–631.

Edmonds, J. (1960) *A Combinatorial Representation for Oriented Polyhedral Surfaces*. URL https://books.google.nl/books?id=vo2ENwAACAAJ.

Elberink, O. (2010) Acquisition of 3D topography: Automated 3D road and building reconstruction using airborne laser scanner data and topographic map. PhD thesis, University of Twente Faculty of Geo-Information and Earth Observation (ITC).

El-Mekawy, M. and Östman, A. (2010) Semantic mapping: An ontology engineering method for integrating building models in IFC and CityGML. *Proceedings of the 3rd ISDE Digital Earth Summit*, pp. 12–14.

El-Mekawy, M., Östman, A. and Shahzad, K. (2012) A unified building model for 3D urban GIS. *ISPRS International Journal of Geo-Information*, 1: 120–145.

Ellul, C. and Altenbuchner, J. (2014) Investigating approaches to improving rendering performance of 3d city models on mobile devices. *Geo-spatial Information Science*, 17(2): 73–84, January. Doi: 10.1080/10095020.2013.866620.

Feng, X., Wang, Y., Weng, Y. and Tong, Y. (2013) Compact combinatorial maps: A volume mesh data structure. *Graphical Models*, 75(3): 149–156, May. Doi: 10.1016/j.gmod.2012.10.001.

Guercke, R., Brenner, C. and Sester, M. (2009) Generalization of 3D City Models as a service. In *ISPRS Workshop on Quality, Scale and Analysis Aspects of City Models*. ISPRS Archives – Volume XXXVIII-2/W11.

Guibas, L. and Stolfi, J. (1985) Primitives for the manipulation of general subdivisions and the computation of opogra. *ACM Transactions on Graphics*, 4(2): 74–123, April. Doi: 10.1145/282918.282923.

Guo, Y., Pan, M., Wang, Z., Qu, H. and Lan, X. (2010) A spatial overlay analysis method for three-dimensional vector polyhedrons. In *2010 18th International Conference on Geoinformatics*. IEEE, June. Doi: 10.1109/geoinformatics.2010.5567674.

Haala, N. and Rothermel, M. (2012) Dense multi-stereo matching for high quality digital elevation models. *Photogrammetrie, Fernerkundung, Geoinformation (PFG)*, 2012(4): 331–343.

Hanson, A. J. and Cross, R. A. (1993) Interactive visualization methods for four dimensions. In *VIS '93 Proceedings of the 4th conference on Visualization '93*, pp. 196–203. ACM.

Hazelton, N. (1998) Some operations requirements for a multi-temporal 4-D GIS. In M. Egenhofer and R. Golledge (eds), *Spatial and Temporal Reasoning in Geographic Information Systems*, pp. 63–73. Oxford University Press.

Hinton, C. H. (1888) *A New Era of Thought*. Swan Sonnenschein & Co.

Horna, S., Damiand, G., Diakité, A. and Meneveaux, D. (2015) Combining geometry, topology and semantics for generic building description and simulations. In F. Biljecki and V. Tourre (eds), *Eurographics Workshop on Urban Data Modelling and Visualisation*. The Eurographics Association. ISBN 978-3-905674-80-4. Doi: 10.2312/udmv.20151343.

ISO (2003) ISO 19107:2003: Geographic information: Spatial schema. International Organization for Standardization, 2003.

ISO (2013) *Industry Foundation Classes (IFC) for Data Sharing in the Construction and Facility Management Industries*. International Organization for Standardization, March 2013.

ISO (2014) *Industrial Automation Systems and Integration – Product Data Representation and Exchange*. International Organization for Standardization, August 2014.

Kada, M. (2008) Generalization of 3D building models for map-like presentations. *The International Archives of the Photogrammetry, Remote Sensing and Spatial Information Sciences: XXXVII.[S. l.]: ISPRS*, pp. 399–404.

Kageyama, A. (2016) A visualization method of four dimensional polytopes by oval display of parallel hyperplane slices. Available at https://arxiv.org/pdf/1607.01102

Kersten, T.P., Keller, F., Saenger, J., Schiewe, J. (2012). Automated Generation of an Historic 4D City Model of Hamburg and Its Visualisation with the GE Engine. In: Ioannides, M., Fritsch, D., Leissner, J., Davies, R., Remondino, F., Caffo, R. (eds) Progress in Cultural Heritage Preservation. EuroMed 2012. Lecture Notes in Computer Science, vol 7616. Springer, Berlin, Heidelberg. https://doi.org/10.1007.

Kraak, M. J. (2008) *Geovisualization and Time: New Opportunities for the Space–Time Cube*, pp. 293–306. Wiley.

Kumar, K. (2017) Integration of CityGML and InfraGML. OGC Technical and Planning Committee Meeting 2017, Delft, The Netherlands. https://3d.bk.tudelft.nl/ kavisha/pdf/ Kavisha_OGC_LandinfraDWG_March212017.pdf.

Labetski, A., Ledoux, H. and Stoter, J. (2017) Generalising 3D buildings from LoD2 to LoD1. In *GISRUK 2017 Conference Proceedings*. URL http://huckg.is/gisruk2017/ GISRUK_2017_paper_92.pdf.

Laurini, R. and Milleret-Raffort, F. (1994) Topological reorganization of inconsistent geographical databases: A step towards their certification. *Computers & Graphics*, 18 (6): 803–813.

Lebesgue, M. (1911) Sur l'invariance du nombre de dimensions d'un espace et sur le theoréme de M. Jordan relatif aux varieté fermées. *Comptes rendus de l'Académie des Sciences*, 152: 841–844.

Ledoux, H. (2013) On the validation of solids represented with the international standards for geographic information. *Computer-aided Civil and Infrastructure Engineering*, 28 (9): 693–706.

Ledoux, H. (2018) Val3dity: Validation of 3D GIS primitives according to the international standards. *Open Geospatial Data, Software and Standards*, 3(1): 1.

Lienhardt, P. (1994) *N*-dimensional generalized combinatorial maps and cellular quasimanifolds. *International Journal of Computational Geometry and Applications*, 4(3): 275–324.

Mallet, C. and Bretar, F. (2009)Full-waveform topographic lidar: State-of-the-art. *ISPRS Journal of Photogrammetry and Remote Sensing*, 64(1): 1–16.

Muller D. E. and Preparata, F. P. (1978) Finding the intersection of two convex topography. *Theoretical Computer Science*, 7(2): 217–236. Doi: 10.1016/0304-3975(78)90051-8.

Nouvel, R., Zirak, M., Coors, V. and Eicker, U. (2017) The influence of data quality on urban heating demand modelling using 3D city models. *Computers, Environment and Urban Systems*, 64: 68–80.

OGC (2007) Geography markup language (GML) encoding standard. Open Geospatial Consortium inc. Document 07-036, version 3.2.1.

OGC (2011) *OpenGIS Implementation Specification for Geographic Information – Simple Feature Access – Part 1: Common Architecture. Version 1.2.1.* Open Geospatial Consortium, May.

OGC (2012) OGC city geography markup language (CityGML) encoding standard. Open Geospatial Consortium inc. Document 12-019, version 2.0.0.

OGC (2016a) AFuture City Pilot-1: Using IFC/CityGML in urban planning engineering report. Open Geospatial Consortium inc. Document OGC 16-097.

OGC (2016b) Future City Pilot-1: Using IFC/CityGML in urban planning engineering report. Open Geospatial Consortium inc. http://docs.opengeospatial.org/per/ 16-097.html.

OGC (2016c) OGC Land and Infrastructure Conceptual Model Standard. OGC Document reference 15-111r1. Open Geospatial Consortium inc. http://docs. Opengeospatial.org/is/15-111r1/15-111r1.html.

OGC (2017) OGC InfraGML 1.0: Part 0 – LandInfra Core – Encoding Standard. OGC Document reference 16-100r2. Open Geospatial Consortium inc. http: //www.opengis.net/doc/standard/infragml/part0/1.0.

Rottensteiner, F. (2003) Automatic generation of high-quality building models from lidar data. *IEEE Computer Graphics and Applications*, 23(6): 42–50.

Shahzad, M. and Zhu, X. X. (2015) Robust reconstruction of building facades for large areas using spaceborne TomoSAR point clouds. *IEEE Transactions on Geoscience and Remote Sensing*, 53(2): 752–769.

Steuer, H., Machl, T., Sindram, M., Liebel, L. and Kolbe, T. H. (2015) *Voluminator—Approximating the volume of 3D buildings to overcome topological errors.* In: Bacao, F., Santos, M., Painho, M. (eds) AGILE 2015. Lecture Notes in Geoinformation and Cartography. Springer, Cham. https://doi.org/10.1007/978-3-319-16787-9_20, pp. 343–362.

van den Brink, L., Stoter, J. and Zlatanova, S. (2013) UML-based approach to developing a CityGML application domain extension. *Transactions in GIS*, 17(6): 920–942.

Vince, A. (1983) Combinatorial maps. *Journal of Combinatorial Theory, Series B*, 34(1): 1–21, February. doi: 10.1016/0095-8956(83)90002-3.

Zlatanova, S. (2000) On 3d topological relationships. In *Database and Expert Systems Applications, 2000. Proceedings. 11th International Workshop On*, pp. 913–919. IEEE, 2000. doi: 10.1109/DEXA.2000.875135.

# Chapter 7
# Updating geometric and thematic information

*David Holland*

## 7.1 Introduction

Ordnance Survey (OS) has been collecting and managing topographic information in Great Britain for more than 200 years. Both field-based and office-based techniques have been used to survey natural and human-made features, surface heights and other associated spatial information such as road routing data, addresses and points of interest. The methods used to capture this information have of course changed over the years, as have the ways in which the information is stored and managed. In this chapter we discuss the methods used to collect data in the field; the photogrammetric and associated workflows; and the ways in which we may collect data in future.

## 7.2 Field data capture

There are still many surveyors at Ordnance Survey (OS) who remember the days when they took a 'sketching case' into the field and drew new features onto a copy of the map using a sharp pencil. Field surveyors would go to the field office at the start of the day, collect the tools they needed for the day's surveys, and go out in pairs to add new buildings, fences, roads and vegetation features to the map. Equipment such as theodolites, distance meters (distos) and, later, total stations were available, but often the surveyor would use more simple equipment such as an optical prism (or 'popeye') to carry out the survey.

One of the major changes to the OS field offices in recent years has been their almost complete disappearance! Most field surveyors now work from home, downloading their jobs over the internet and working largely on their own, apart from the occasional meeting with their colleagues at a team or regional get-together. The kit now consists of a rugged tablet computer, a GNSS receiver (and a pole). Instead of fixing their position with reference to known positions (such as the famous 'trig pillars' located at vantage points all over the country), the GNSS kit uses a national network of reference stations (OS Net) which provides the corrections necessary to give precise coordinates, to an absolute accuracy of a few centimetres. At the moment, OS Net comprises more than 100 stations, located throughout Great Britain to provide accurate coordinates anywhere in the country. The station coordinates are known to a high degree of accuracy and are used to determine the transformation from 'GPS coordinates' (or, more accurately, ETRS89 coordinates) to the British National Grid coordinates that OS still uses for its maps and spatial data products. The latest version of this transformation, OSTN15, is available as a software package for users to download, or as a set of parameters and algorithms that developers can incorporate into their own software.

### 7.2.1 Changing data requirements

Many of the features on the early OS maps were there because of their significance in a military context (hence the name 'Ordnance Survey'). Today, the users of OS data are from a wide range of industries and organisations, including utility companies, central and local government departments and agencies, and general consumers. The spatial data are now put to a wide range of uses, and the list is growing all the time. Future users of national mapping data are likely to include smart city architects; autonomous vehicle manufacturers; and 5G telecommunications planners. Perhaps these will become the major users of spatial data, but new potential users are appearing all the time, so we have to be able to adapt to the changing requirements while retaining the information required by our traditional users.

The geometric accuracy of the topographic data used to be the main consideration when assessing its quality and fitness for purpose. Finding new methods to capture data of high geometric accuracy is still of importance, but other factors now come into play, such as the currency of the data and the non-spatial attributes of the geospatial features. The data currency could be the date at which a feature was updated, but for some users it is equally important to know when the feature was last checked, to see if any changes had occurred. Although this may not lead to any material change to the spatial data (i.e. if there has been no change) it will change the metadata (i.e. the date on which the feature was last checked), which is of equal value in some applications. The non-spatial attributes of a building feature might include such things as the year in which it was built; the name of the business occupying the property; the number of storeys it has; or the shape of its roof.

One aspect of spatial data that has been investigated repeatedly over the years is the capture and maintenance of three-dimensional data. The main thing we have learned from this research is that there are as many different interpretations of 3D data as there are potential users. While a grid-based digital surface model may be fine for one user, another may require a coloured point cloud; a third may want a textured mesh model, while yet another will demand a fully-fledged 3D city model with all the walls, roofs and other structures modelled explicitly. Which of these is the right model to use depends on several factors, including: the use cases in which it will be applied; the software, hardware and storage facilities available; the amount of processing, and time, required to produce the data; the amount of analytical processing that will be applied to the data; and, not least, the amount of money the user is willing to pay. This chapter describes some of the methods Ordnance Survey has used to capture geospatial data, and the research it has undertaken into the data capture and data processing methods it is likely to use in future.

### 7.2.2 Depicting 3D data

Ordnance Survey has depicted height information in its products right from the first maps of the early 19th century. In 1801, the OS map of the county of Kent used a shading technique, known as hachuring, to indicate areas of sloping terrain, as shown

Figure 7.1 An example of the hachuring process, indicating slopes on early maps, such as this map of Kent from 1801.

in Figure 7.1. Later, contours were introduced to provide a more quantified depiction of height. From a contour map it is easy to calculate the ground height at any point, but it is not always easy to visualise the shape of the landscape by looking at the contours. To make the visualisation process easier, colouring and shading have sometimes been used to give an added cue to the contours. Shading similar to the original hachuring gives an impression of the shape of hills and the shadows they cast. By convention, the shadows are usually depicted as if the sun is in the north west – so the shadows are only an indication of form, rather than a depiction of where real shadows actually fall.

When OS began to convert from map-based to digital data, the contours in the maps were digitised as vectors, with the contour height as an attribute. This form of digital height data is difficult to work with and it was soon replaced in popularity by digital terrain models (DTMs), derived originally from the contours and other height data on the map (spot heights captured either in the field or using photogrammetric techniques). Gridded digital terrain models are easy to work with and can be interpreted in a similar way to image data, where the pixel values represent height rather than colour. A DTM can also be simply treated as a matrix, making it easy to analyse in both GIS and mathematical analysis software.

In recent years, other ways of depicting 3D data have appeared, often connected with the method by which the data were captured. For example, an aerial laser scanning (LiDAR) system produces a dataset consisting of millions of points in 3D space, which can most easily be depicted as a point cloud. The colour of each point may be derived from the intensity of the laser return signal; the height of the point above a reference frame such as sea level; or a colour interpolated from an optical image captured at the same time as the LiDAR.

At OS, point clouds are created not from LiDAR data but from the aerial imagery captured originally to update the topographic map data. The image-matching technologies used to derive point clouds from multiple-overlapping images have vastly improved over the years. In the early days, image matching produced a point cloud which, when visualised, resembled a scene covered in thick snow, where all the detail was smoothed over, especially at the edges of buildings, trees and other features. Current image-matching software produces point clouds that approach LiDAR point clouds in quality. These point clouds can be coloured from the imagery used to produce them, giving a dataset with attributes of both an image and a 3D model. Figure 7.2 shows an example of a point cloud derived from aerial imagery. Although point clouds created in this way have been shown to be sufficient for Ordnance Survey's requirements, it should be noted that they are not identical to those produced by LiDAR. One difference is that LiDAR points are measured values, while image-matched points are inferred from the images and the geometry of the capture scenario. This difference is not always apparent in the resulting point cloud and, for the level of accuracy required for a topographic mapping survey, is usually not an issue. Where LiDAR data differ significantly from image-matched point clouds is in areas of forest. Here, the narrow laser beam of a LiDAR system often penetrates through foliage to reach lower parts of the vegetation and the ground beneath, while an aerial image will only portray the

Figure 7.2 A point cloud derived from aerial imagery.

top of the forest canopy. In practice, this means that LiDAR systems are more capable of measuring the structure of the vegetation and the ground beneath it. Up to now, Ordnance Survey has found that this advantage of LiDAR has not been enough to justify its use in areas where we already capture multi-overlapping imagery.

### 7.2.3 Processing flowline

The system used at OS in 2018 is based on Vexcel UltraCam XP cameras, processed using Vexcel UltraMap software. Camera data are processed to derive a set of outputs: oriented frame images for use in stereo-viewing software; orthorectified colour-balanced images for use as an image product; digital surface models and digital terrain models; and data for use in image classification and change detection processes. Two Cessna 404 aircraft (Figure 7.3) are deployed from February to November, collecting images all over Great Britain, at a rate of over 50,000 square kilometres per year. Images are usually collected as blocks, each block covering an area 20 km by 10 km. The choice of where to fly is governed by the need to update an area; the presence of clear skies; and permission to fly in suitable airspace, obtained from the Civil Aviation Authority. These various factors do not always align with each other – beautiful clear skies over a major city will often not coincide with permission to fly over that city; while permission to fly over certain Scottish islands may be much easier to obtain but the absence of clouds in that area is far less easy to arrange.

The images are captured at a flying height and spacing to allow the production of images with 15 cm ground sample distance (pixel size on the ground) and 60% to 80% fore/aft overlap. Flight lines are planned to allow side overlap of at least 30%.

Figure 7.3 One of the OS survey aircraft.

Once captured, the images are sent to OS headquarters in Southampton for processing. Vexcel UltraMap is used to carry out the processes of aerial triangulation, orthorectification and digital surface model creation. An archive of ground control points, augmented by newly captured ground survey when necessary, is used in the aerial triangulation process to ensure that the imagery meets the geometric accuracy required. The results of aerial triangulation are fed into the topographic data capture process, in which photo-interpreters use stereo-viewing techniques to update the large-scale topographic data from the images. A parallel flowline, run by external data capture companies, uses orthorectified imagery to achieve a similar result.

The software used to view and capture vectors from the stereo imagery has varied over the years but is currently based on DAT/EM's Summit Evolution photogrammetric workstation software and ESRI's ArcGIS, tailored to fit OS's data capture requirements. The same ArcGIS software is also used by surveyors to capture topographic data in the field.

In addition to the topographic vector data, the imagery is used to create a separate image product (OS MasterMap Imagery Layer) and a digital terrain model (OS Terrain 5). The image product is designed to provide a mosaic over the whole country, with adjacent frames colour-balanced and merged to give, where possible, a homogeneous look to the entire scene being viewed. A rather different image product is created for use in internal image classification projects. This second image product preserves the original radiometry and retains all four bands (blue, green, red and near infra-red) of the images. The classification process which utilises this imagery is described below.

## 7.3 Change detection and classification

Automatic change detection has long been a goal of national mapping organisations. In order to update topographic and associated data, a mapping agency has to know where and when changes to the landscape have occurred. This information may come from third parties such as local administrations and building contractors, or by direct observation by field surveyors, or may be picked up visually by photogrammetrists as they view the stereo imagery or ortho-photos. The process of finding the changes is quite time consuming and therefore expensive, so any method of enhancing the process is welcome. The OS research team has been investigating methods of automatic change detection for many years (Holland *et al.*, 2012), and has developed a system that is now used operationally. Having tested different methods of change detection, including image-to-image and polygon-to-polygon, a system has been developed based on the comparison of a classified image with the contents of the existing topographic database.

The first stage in the change detection flowline is the generation of the input data. Imagery from the aerial camera is processed to create an orthorectified 4-band image, at the original resolution of the sensor, and without any radiometric processing applied. A second input is the digital surface model (DSM), produced from the original imagery using image-matching techniques (in UltraMap software).

A digital terrain model (already in existence, or derived from the DSM) is also used. From these inputs, various other inputs are created, such as the normalised DSM (the DSM minus the underlying DTM, to give the heights of all objects above the surface) and a slope map. All of these inputs are entered into Trimble's eCognition classification software.

The software first segments the image into objects, then uses the radiometric and height data to derive a class for each of these objects. This uses a rule-based system which takes into account not only the input data but also the characteristics (size, shape, texture, etc.) of the segments and the neighbourhoods in which they occur. In our current system, an extensive ruleset has been built up to characterise seven separate classes: low vegetation (grass/crops), medium vegetation (scrub), high vegetation (trees), buildings, sealed surfaces, unsealed surfaces, and water. The process of assigning an object to one of these classes can be quite complex and has been built up over several years by experts in image interpretation at OS. Later, we will discuss possible alternative techniques that remove the need for such an explicit set of rules. After classification, the resulting objects are compared with the contents of the existing large-scale topographic database. Any differences are reported as potential changes, termed 'change candidates'. For example, an object in the image classified as a building, in the same space as an area of vegetation in the topographic database, will be flagged up as potentially a new building. Each change candidate is then passed on to the photogrammetrists, who verify whether or not it is a real change and, if it is, capture the new information into the topographic database. As there is always a human being at the end of the process to verify the data, we have tuned it to find as many of the real changes as possible, at the expense of a large proportion of false positives. The photogrammetrists can rapidly eliminate these false positives, so false positives only take up a small amount of extra time. Compared with the previous system, where the photogrammetrists had to find and identify all the changes themselves, we have found that automatic change detection makes for a more efficient and effective workflow. Figure 7.4 shows an example of the classification process.

As the first stage in change detection is a classification of the images, it became clear to us that this in itself could be a useful addition to our products. At the time of writing, the classification flowline has been developed specifically to identify hedges, for use in the verification of EU Common Agricultural Policy subsidy applications. Similar modifications could in future be used to classify trees, water, or other land cover types.

## 7.4 Machine learning

As noted earlier, the development of rule-based classification systems can be a long and complex process. Over recent years, the development of machine learning techniques has led to new ways of classifying images that do not depend on rules but, rather, only on matching the input data with the required output. Much of the development of machine learning from images has been on the identification of objects within terrestrial photos. Today, machines can identify cups, cars, tigers, bicycles and

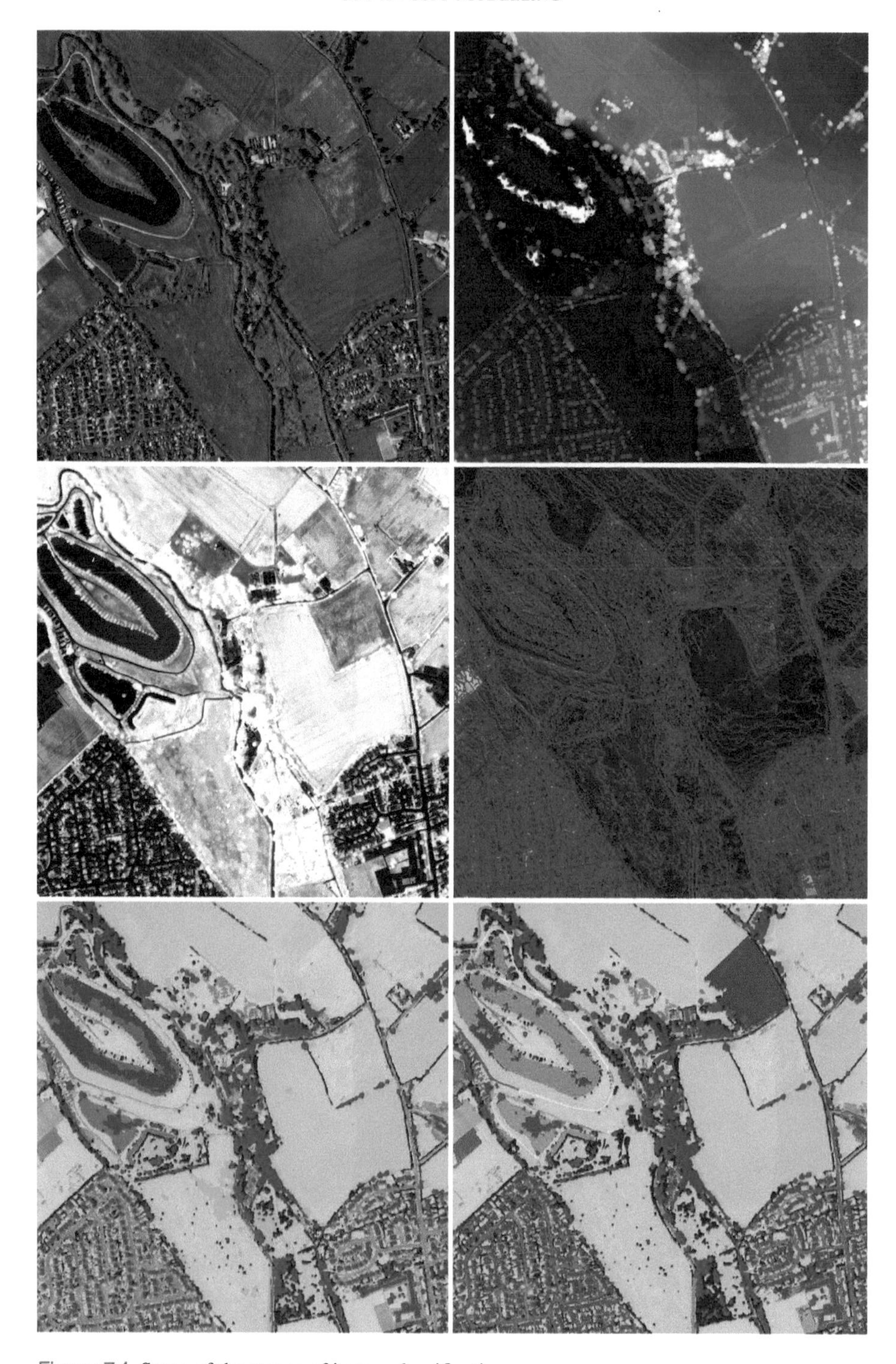

Figure 7.4 Some of the stages of image classification.

many other objects in images with a greater success rate than the average human. For a mapping agency, it would seem natural to attempt to develop similar techniques to identify objects in aerial images. There are differences in the two scenarios, however. In the traditional case, such as the objects in the imageNet dataset (Deng *et al.*, 2009), most are pictures of an object within a neutral scene. Thus, there is a foreground object which the machine learning algorithm identifies, and a background which generally plays no part in the identification process. With aerial images, the situation is somewhat different: almost everything in the image (tree, house, road, lake, field, etc.) could be taken as either foreground object or as background. We really would like to identify everything in the image, whether it is a woodland or a building sited within it. The challenge is to find a way to identify all these foreground/background objects, or at least to find the ones we are most interested in. In order for machine learning networks to be trained, a large database of labelled objects is required. Online research datasets exist in some domains: for example, for natural images of objects the ImageNet database can be used, while for number recognition the MNIST dataset is often used. For aerial images, one source of labelling that we already have is the large-scale topographic polygon data of OS MasterMap, in which each feature will normally match an equivalent object in an aerial image. We could start by training a deep learning network to identify parts of the image that are labelled as buildings, vegetation, water or sealed surfaces in the mapping data. In a collaboration between Ordnance Survey and the University of Southampton, this is exactly what was tried. A set of image patches, labelled with OS MasterMap polygon labels, was trained using a convolutional neural network. The results were far from perfect but were promising enough to initiate a project to investigate deep learning techniques further. Our ultimate goal is to be able to 'decode the landscape' by finding patterns in the urban and rural landscape which represent features and concepts not yet captured in the topographic data. For example, the road layout of a housing estate could indicate the decade in which it was built, while the shape and size of buildings might indicate an industrial rather than a residential area. Further work in this area is now underway, to determine how best to utilise machine learning within a mapping organisation.

## 7.5 New sensors and platforms

There has been a rapid evolution in the world of sensing devices over the past decade. This can readily be seen in the consumer market, where high-end camera technology has made its way into almost everyone's pocket in the shape of a smart phone (are there many mobile phones left that are not 'smart'?). Ten years ago mapping agencies were only just making the change from wet film technology to aerial digital cameras but, since then, the capability of such cameras has grown gradually, while the capability of consumer-grade cameras has also grown rapidly. The light-gathering ability of sensors has improved, the number of pixels per image has grown, and the speed of capture (frames per second) has increased. In a similar way, other technologies once associated with the surveying profession have moved into the mainstream. GPS and other global navigation satellite systems (GNSS) once required specialist hardware and software to enable the user to obtain a high-quality measurement of position.

GNSS devices have gradually reduced in size and increased in performance, with the smart phone again being the vehicle that has taken such systems to the mass market.

At Ordnance Survey, a team of researchers continues to investigate the changing technology, in order to find the best solution for our future data-collection needs. This, of course, involves research into new survey equipment, mobile mapping systems and aerial cameras, but it also takes into account the development of less specialised equipment. One such investigation involved the use of unmanned aerial vehicles (UAVs).

Although OS has had use of a UAV for several years, the model we first used proved to be unsuitable for our needs in Great Britain. This Sensefly Swinglet could be used to capture images over a predetermined flight plan but was susceptible to any but the lightest of winds. The method of landing involved shutting off the power to the propeller and the UAV gliding to the ground. This left the operator with no control over the descent, which could in some circumstances lead to the aircraft being blown off course and landing in awkward places (trees and bushes being particular favourites). In 2016 it was decided to test out a more robust, survey-grade UAV to evaluate its potential as a survey tool for use in field operations. The platform we chose was an Aibot X6 from Hexagon Geosytems (Figure 7.5). This is a hexacopter with a gimbal mount onto which a variety of cameras may be fixed. We tested several camera models, settling on a Sony A6000 mirrorless compact camera, with a 20 mm lens.

Figure 7.5 The Aibot X6 UAV used by Ordnance Survey.

A team of three UAV pilots went through the training and certification process, to obtain Civil Aviation Authority permission for commercial operations. Once this was granted, the team worked on a project to test the UAV on various potential use cases, including the capture of solar farms, quarries, a holiday park and other areas where a UAV might have an advantage over conventional techniques. Traditionally, surveyors will find such areas of change which they are unable to capture by field survey, for reasons such as difficulty of access, health and safety concerns, or the complexity of the change in question. These areas are reported to the OS flying team, who schedule them in to be captured by aerial survey in between other jobs or, in certain circumstances, as a stand-alone flight. As these areas are often quite small, it can be quite expensive to send out the conventional aircraft to capture what may be only a handful of images. It is these types of jobs that we considered to be the most suitable for flying with a UAV. The data-capture phase of the UAV tests included the planning of the flights, obtaining permission from the relevant landowner, informing the local air traffic control when necessary, choosing a suitable day (and suitable weather) for flying, and then flying the UAV and capturing the images. After this, the images were processed using Pix4D to obtain a digital surface model and an orthorectified image of each area. The ortho-photos were then used as the basis of topographic vector capture using the same ArcGIS system as used in production. Figure 7.6 shows a 3D model of the Ordnance Survey office, created from UAV imagery in Pix4D.

After several months of testing, a cost-benefit analysis was carried out to determine whether or not the UAV had proved its worth as a data collection tool. We were pleased to find that our work had not been in vain and that the UAV could be a viable tool for use in the field, in certain circumstances. The main uses for a UAV were found to be in sites of around 8–10 hectares, especially where there was a health and safety issue for field surveyors. Suitable areas included construction sites, quarries, and sites where the ground surface is unstable or dangerous. These findings were accepted by

Figure 7.6 A 3D mesh model of the OS office in Southampton (Explorer House), created from UAV imagery.

OS and we have now embarked on a full-scale field trial, in which a field surveyor will be trained to fly a UAV and will use it on a variety of jobs in one of the regions of the UK. If this proves successful, it may be extended to multiple UAVs over the rest of the country.

## 7.6 Crowdsourcing

One data source that OS has not exploited to any great extent is that of volunteered geographic information, or crowdsourcing. It could be seen as obvious that members of the public should be able to report changes to landscape features that have not yet appeared in OS data or products. There have been several barriers to this, not least that of how to incentivise the process to encourage people to participate. When there is an obvious advantage to taking part in crowdsourcing activity – such as the provision of new and up-to-date data within an open dataset – it can be quite easy to encourage participation. However, for a national mapping agency that is also a commercial organisation – charging customers for its products – it can be more difficult to motivate the public to provide free data. Notwithstanding this, crowdsourcing is such a potentially important method of data capture that a mapping agency cannot afford to ignore it.

In order to familiarise ourselves with crowdsourcing technology, a project was instigated to investigate methods used elsewhere, and to develop a small-scale crowdsourcing project using what we hoped would be a 'friendly crowd' of OS employees. After several potential platforms had been investigated, we settled on the Zooniverse platform, which is well established in the crowdsourcing world (https://www.zooniverse.org/). Several Zooniverse projects were set up, in which OS staff were asked to spend a few minutes identifying features in images. These ranged from finding lamp-posts in mobile mapping data to classifying roof types in aerial photography. This type of crowdsourcing proved to be a quick way to capture a large sample of simple information, to be used in other processes (e.g. labelling test data for machine learning). Developing a system for internal use is only a small step, but it did provide us with the ability to build and deploy a useful crowdsourcing application. Work is planned to develop more internal applications, and to further investigate the potential of external crowdsourcing as a means of data collection.

## 7.7 Summary

This chapter has described some of the recent work undertaken by Ordnance Survey in the pursuit of more efficient geospatial data collection and processing. Some of the techniques described will become part of standard working practice in the near future; others may prove to be too expensive or not efficient enough to be viable in a production environment. One thing we have learnt is that the pace of change continues to accelerate, with new technology and new processing algorithms appearing all the time. And it seems to be the case that the speed of development is also increasing, with new research techniques rapidly being assimilated into the mainstream. This can make it difficult for a small organisation to keep up with the change, but Ordnance Survey has shown that, by focussing on the data and products it requires, and by targeting the technologies it believes are the most promising, it can develop solutions to our current and future data collection and processing requirements.

## References

Deng, J., Dong, W., Socher, R., Li, L.-J., Li, K. and Fei-Fei, L. (2009) ImageNet: A large-scale hierarchical image database. *IEEE Computer Society Conference on Computer Vision and Pattern Recognition (CVPR 2009)*, 20–25 June 2009, Miami, Florida, USA

Holland, D., Gladstone, C. and Gardiner, A. (2012) A semi-automated method for detecting changes to Ordnance Survey topographic data in rural environments. Proceedings of the 4th GEOBIA, May 7–9, 2012, Rio de Janeiro, Brazil.

# Chapter 8
# Change detection for geodatabase updating

*Rongjun Qin*

The geodatabase (vectorized data) is nowadays very much a standard digital city infrastructure; however, updating a geodatabase efficiently and economically remains a fundamental and practical issue in the geospatial industry. The cost of building a geodatabase is extremely high and is labor intensive, and very often the maps we use have several months and even years of latency. One solution is to develop more automated methods for (vectorized) geospatial data generation, which has been proven a difficult task. An alternative solution is first to detect the differences between the new data and the existing geospatial data, and then only update the area identified as changes. A second approach, that is becoming more favored due to its high practicality and flexibility, is the highly relevant technique of change detection. This chapter aims to provide an overview of state-of-the-art change detection methods in the field of remote sensing and geomatics to support the task of updating geodatabases. Data used for change detection are highly disparate and, accordingly, we structure this review intuitively based on the data being (1) change detection with 2D data; or (2) change detection with 3D data. Conclusions are drawn based on the review together an outlook is provided of the topic of updating geodatabases.

## 8.1 Introduction

### 8.1.1 Background

The world never remains static. The moment a geodatabase (vector data) is constructed (normally at a high cost), a substantial practical problem follows: How can these data be updated economically and in a timely fashion given the rapid development in land cover change and the dynamics of man-made objects. In many cases, the maps (geodatabase) we use have at least several months of latency. However, current approaches – for example, updating the basemap of a car navigation system – rely on hiring thousands of people to drive every street/road in the urban/suburban/rural area to manually record, verify and regenerate the maps, which is extremely time consuming and expensive. Intuitively this points to a solution that approaches the construction of geodatabases in a more automated way. However, this is a long-standing research problem that requires a technical breakthrough in automated feature extraction, object extraction and topological reconstruction; unfortunately, these problems have remained largely unresolved for many decades (in terms of application of engineering-grade data generation). An alternative solution – change detection – proposes not focusing on database construction speed and automation but, rather, that we identify potential changed areas efficiently and only update them. This is apparently related to developments in the well-known change detection problem, which performs a direct comparison of multi-temporal data to yield temporal differences for geodatabase updating. The change detection and updating approach (sometimes called change modelling, which is considered identical to change detection in this chapter) is becoming a preferred method in addressing the geodatabase updating problem, mainly due to several advantages:

1. In comparison to the approach of developing more automated methods in geospatial data extraction, change detection approach can significantly reduce the amount of data being remodeled and has the potential to be deployed in a relatively short time.
2. Change detection algorithms normally aim to automatically detect the differences in data; this reflects a much faster identification of the dynamics of ground objects, and the change itself can be used as an independent product.
3. There are plenty of timely (and sometimes free) multi-temporal data sources available, with levels of detail ranging in scale from continental to city and individual object, allowing development of flexible multi-resolution strategies for change detection to further reduce the cost.

### 8.1.2 Scope of this chapter

Change detection (CD) in remote sensing and geomatics refers to the process of identifying differences in the state of the interested object or phenomenon by observing it at different times (Singh, 1989). It is the most important step for updating the geodatabase, where the updating itself is followed by a reconstruction (vectorization) of the changed object, which uses common approaches (manual or semi-automated). This chapter provides an overview of start-of-the-art approaches of CD and geodatabase updating. CD is a well-defined and classic topic in the field of remote sensing, with the aim of detecting the differences and monitoring the land process and dynamics at a relatively coarse level. The focus is primarily on methods that adopt remote sensing and geomatics data for geodatabase updating at various levels of detail. Other relevant research topics, such as real-time video sequence analysis, multi-temporal data analysis and indoor/outdoor object tracking, partially regarded as a process of change detection, are beyond the scope of this review. We primarily consider change detection approaches in bi-temporal data (before and after), and methods using multi-temporal data that will be occasionally presented.

The types and formats of data may vary greatly, and there are generally two types of data: raster data and vector data. To be specific, raster data are unstructured raw data that come directly from the sensors or only through preliminary low-level processing (e.g. de-noising, format conversion, decoding), and examples are images, stereo/multi-stereo image blocks, filtered/raw 3D point clouds, and 3D triangular meshes. Vector data refers to structured data where necessary information has been extracted, such as DLG (digital line graphs), 3D polyhedral models, building footprints with associated information, e.g. the number of floors of the building, area/type of the buildings. Updating vector data using images/3D point clouds is directly relevant, while methods that use purely bi-temporal images/point clouds are equally relevant, as the extracted changes can be directly applied to the geodatabase for updating. In providing a review of change detection methods in the field of remote sensing and geomatics, this chapter is limited to methods that use (1) unstructured raw data: optical images, LiDAR (light detection and ranging), point clouds, triangular photo meshes from airborne, satellite and terrestrial platforms; (2) structured data: DLG, 3D polyhedral models (LoD models). Other data types, such as sonar, microwave radar, and ground penetrating radars, are not included.

### 8.1.3 Organization of this chapter

CD is a highly disparate problem that varies greatly with data quality, resolution and format. This leads to many different types of categorization of CD methods, e.g. based on the basic processing unit, data types and processing methods (Lu *et al.*, 2004; Tewkesbury *et al.*, 2015). This chapter adopts a very intuitive classification of methods, based on the dimension of the data, i.e. (1) change detection with 2D data, and (2) change detection with 3D data. Section 8.2 briefly introduces the necessary data preprocessing methods. These two types of method are examined in Section 8.3, Change detection with 2D data, and Section 8.4, Change detection with 3D data, based on the predominant methods in the existing literature. In the final section, we share our concluding remarks and our outlook on the topic of geodatabase updating. It should be noted that 2D and 3D CD are not completely isolated; methods that use a mixture of 2D and 3D data (e.g. using DSM to update DLG), we categorize as change detection with 3D data.

## 8.2 Data preprocessing

Raw data acquired from different platforms need to be processed into a geo-referenced/formatted dataset. Geo-referencing of data acquired from different platforms vary due to the data formation and accuracy of the onboard positioning sensors (i.e. GPS (Global positioning System)/IMU (Inertial Measurement Unit)).The geo-referencing of images and LiDAR nowadays has become a routine procedure, with many processing software packages available (HEXAGON, 2014; Pix4D, 2017; Terrasolid, 2013). Satellite data are often coarsely geo-referenced based on the positioning sensors in the spaceborne platform prior to their release. The unstructured data contain images and point clouds, generically representing spectral (colour) and geometric information (2D/3D shape). Therefore, before CD, data alignment is needed on both spectrum and geometry, being radiometric correction and geometric co-registration. For CD with stereo/multi-stereo images, a DSM generation stage is normally necessary to process the 3D information.

*Radiometric correction:* Radiometric correction has been a basic process in many image-based CD methods. It can generally be categorized into absolute and relative methods (Hu *et al.*, 2011). The absolute method aims to recover the absolute surface reflectance (Gordon, 1997) based on the radiometric values from the images, while relative correction converts the spectrum (colour) of an image to a reference image (sometimes of the same scene) (Roy *et al.*, 2008).

Absolute methods are normally used when dealing with low-to-medium resolution (LTMR) remote sensing data, usually through a process of atmospheric correction using radiative transferring models (Berk *et al.*, 1999). The recovered surface reflectance is seen as invariant of sensors and acquisition conditions, such that they can be used for direct comparisons. However, to accurately recover the surface reflectance, many in-situ data are necessary, such as weather, aerosol optical depth, humidity and so on, which are often hard to obtain. Therefore, absolute methods are only considered when identification of the actual values of surface reflectance is needed, or when they have already been processed through other applications. However, for the purpose of updating geodatabases, the actual surface reflectance values are not necessary.

Practically, relative methods (relative correction or relative normalization) are preferred as they do not require additional observations except for the images themselves. This usually refers to the correction of a multiplicative and additive intensity change, and the correction parameters can be estimated either through a few reference pixels or all pixels and patches (El Hajj *et al.*, 2008; Yang and Lo, 2000). The maximum-and-minimum range normalization does not consider the additive intensity change (Yang and Lo, 2000). In the process of estimating the correction parameters (scale and offset, relevant to multiplicative and additive intensity change), blunder pixels such as those saturated by clouds or significantly changed areas should be eliminated (Paolini *et al.*, 2006). One of the recent methods uses spatio-temporal filtering to perform pixel-wise correction (Qin *et al.*, 2016b). This approach applies non-parametric correction to multi-temporal datasets using a 3D bilateral filter. This approach can further be applied to bi-temporal images; however, it works more effectively with multi-temporal images.

*Data co-registration:* Co-registration refers to the process of aligning the dataset into the same coordinate frame, such that the data from corresponding geographical locations can be compared directly (Chen *et al.*, 2014; Fung, 1990; Fung and Ledrew, 1987). The key for data co-registration is to find sparse or dense corresponding points/areas (e.g. the whole dataset), for applying geometric transformations between two datasets. The correspondences can either be selected manually or extracted using automatic algorithms (Rogan *et al.*, 2002). The co-registration of 2D data assumes that both 2D datasets represent planar spaces, where affine transformation or projective transformation (also called homography) (Hartley and Zisserman, 2003) can be applied. Such approaches work with orthorectified 2D data (parallel projection); however, when applied to perspective images, it requires images taken from the same perspective, or nearly nadir for remote sensing images (Bouziani *et al.*, 2010; Pacifici *et al.*, 2009). Moreover, the height relief of the ground objects cannot be ignored for high-resolution images and affine assumptions are no longer suitable (Qin, 2014a). A compromise solution would be correcting the distortions using TIN (Triangulated Irregular Network), which is based on sparse (corresponding) points between images (Qin *et al.*, 2013; Wu *et al.*, 2012).

Very often remote sensing images are coarsely geo-referenced; therefore they align approximately to a geo-coordinate system. When these images are applied to update the DLG, there are two intuitive approaches:

1. Find correspondences by applying different corner extraction methods on both the DLG (vector data) and image data points (either manually or automatically selected) and then apply rigid, similarity or affine transformations (Zang and Zhou, 2007).
2. Perform an object detection algorithm to extract object footprints (e.g. buildings), and then apply binary image matching and/or vector data matching (Bouziani *et al.*, 2010). This often requires the images to be orthorectified to a digital elevation model or at least mean sea level to correct at least part of the project distortions.

Performing co-registration between 3D datasets is advantageous, since usually the 3D data alignment can be well modeled by 3D rigid or similarity transformations. However, the types of 3D data may bring complicated scenarios: there are normally two types of 3D data. These are:

a. data carrying explicit 3D information such as DSM, 3D point clouds, 3D models, termed 3D-EXP data; and
b. data carrying implicit information (termed 3D-IMP), such as stereo/multi-stereo images that have potential to generate 3D information (Qin *et al.*, 2016a).

Depending on the input bi-temporal data pair (3D-EXP, 3D-IMP or their mixture), the co-registration can be applied either using the imaging sensory geometry (Fischler and Bolles, 1981) or direct 3D transformations. A common approach of co-registering two sets of 3D-IMP data or mixture of 3D-IMP and 3D-EXP is to use a set of GCPs (ground control points) or corresponding 3D/2D feature points, through the process of relative orientation or bundle adjustment (Fraser and Hanley, 2003; Triggs *et al.*, 2000). In particular, if a large number of 2D image correspondences are used under a rigorous sensory model, high-accuracy data alignment can be achieved (Qin, 2014b) for bi-temporal and multi-temporal datasets (Qin *et al.*, 2016*b*). It is normally recommended to co-register two 3D-IMP datasets before converting them to 3D-EXP datasets, since the process of generating 3D-EXP data from 3D-IMP data (e.g. DSM generation from image blocks) may produce errors (Qin, 2014a, 2014b; Qin and Gruen, 2014).

Methods for co-registering two sets of 3D-EXP data (e.g. 3D point clouds, DSM, triangular meshes) are normally performed through 3D rigid or similarity transformations, using either a set of sparse corresponding points or the whole dataset (global method). Global methods minimize the summed squared error of all the points of two datasets, such as least squares 3D matching (Gruen and Akca, 2005) and iterative closest point (ICP) algorithm (Besl and Mckay, 1992). These global methods have outlier removal procedures that are robust to data with a certain level of noise. The co-registration of DSM is usually simplified by estimating a 3D shift between two datasets, while terrestrial 3D-EXP data are often more complicated for co-registration due to the complex geometry and occlusions, which might require good approximation values when both datasets are in different coordinate frames.

*DSM generation*: A necessary step for converting oriented stereo/multi-stereo images into DSMs or point clouds is to perform dense image matching (DIM). DIM methods with multiple images can generally be categorized based on how images are structured (Remondino *et al.*, 2014): (1) multi-stereo matching with depth fusion (MSM); and (2) multi-view matching (MVM). MSM is a direct extension of two-view stereo matching, in which images are paired and point clouds of each pair are fused/filtered (in the depth direction) to form final point clouds (Haala and Rothermel, 2012a; Hirschmüller, 2005). MVM considers matching points across multiple images simultaneously (Baltsavias, 1991; Furukawa and Ponce, 2010). MVM is a more rigorous way to incorporate redundant information but is often more complicated to implement. Both methods have advantages and disadvantages, and their performances vary with the camera network, scene content and complexity, and strategies for point matching (global or local) (Tao *et al.*, 2001; Yang, 2012). Our own experience is that generally, for top-view photogrammetric image blocks (60–80% overlap for frame images and 15–25 degrees intersection angle for satellite images), MSM methods such as those using SGM (semi-global matching) appear to be a good choice, leveraging both speed and performance (d'Angelo and Reinartz, 2011; Krauss *et al.*, 2013). However, for terrestrial images, especially for those that form large baselines and poor camera networks, MVM methods in general produce more complete point clouds, since the visibility is modeled while many stereo algorithms tend to resist objects with large parallaxes (Morgan *et al.*, 2010; Seitz *et al.*, 2006).

## 8.3 Change detection with 2D data

In the remote sensing domain, image-based 2D change detection (CD) has been intensively investigated in the past with the many review papers (Coppin *et al.*, 2004; Hecheltjen *et al.*, 2014; Hussain *et al.*, 2013; Jianya *et al.*, 2008; Lu *et al.*, 2004; Radke *et al.*, 2005; Singh, 1986, 1989; Tewkesbury *et al.*, 2015), and even reviews of review papers, covering each stage of the development of CD methods (İlsever and Unsalan, 2012). Image-based 2D CD algorithms are essentially based on image analysis techniques. They aim at detecting the appearance, disappearance and change of the ground objects by analyzing and interpreting the bi-temporal remote sensing images. In most cases, such techniques are applied to images for small-scale change analysis such as urban sprawl and deforestation using LTMR images, with only a small number of them using VHR images. In general, 2D image-based CD approaches can be categorized into three groups: (1) pixel-based methods; (2) object-based methods; and (3) classification-based methods. The first two categories are distinguished by the basic unit for change representation. The classification-based approaches could be either pixel-based or object-based, but the units are interpreted by performing a pre-classification or post-classification analysis. Methods that use images to directly update the vector databases may fall into any of the three categories, and relevant work will be introduced in each of the three categories.

### 8.3.1 Pixel-based approaches

Pixel-based CD refers to methods that use pixels as the basic processing unit to check the consistency between two datasets for CD. This class of methods is mostly applied to LTMR images (e.g. Landsat, MODIS, Sentinel), since one pixel is significant enough to represent the whole or a relatively large part of the ground, and the mixed-pixel effect (Bioucas-Dias *et al.*, 2012) in this case makes them more robust (less spectrum variation) for CD in a landscape level (Lambin and Ehrlich, 1997; Lu *et al.*, 2002; Mas, 1999; Metternicht, 1999; Ram and Kolarkar, 1993).

Most of the existing CD methods on LTMR images analyze the discrepancy between image spectral information as well as their transformations. Early works include visual checking with image overlay, image differencing (Ingram *et al.*, 1981; Jensen and Toll, 1982), image ratio (Howarth and Wickware, 1981; Nelson, 1983), change vector analysis (Johnson and Kasischke, 1998; Malila, 1980), background subtraction, and transformation-based methods such as principal component analysis (PCA) (Fung and Ledrew, 1987; Singh, 1989), tasseled cap (TC) (Haverkamp and Poulsen, 2003), etc. These methods assume that changes in the land-cover/land use must result in differences in their radiance values, and these differences should be larger than those caused by other factors such as differences of atmospheric condition, surface moisture and sun angles. This assumption works for the LTMR images in most cases, as these variations are saturated by the mixing pixel effects. In the following paragraph, we introduce the predominant methods in pixel-based CD methods.

Image differencing refers to the idea of simply subtracting two geo-referenced images with their radiometric values (Muchoney and Haack, 1994; Singh, 1986). It requires preprocessing steps such as absolute/relative radiometric correction and pixel-wise co-registration of the images. After the image differencing, a binary difference map will be generated by applying a pre-defined threshold to the differences, determined through either trial-and-error approaches or statistical measures based on the standard deviation and mean of the image differences. The image ratio computes the fraction between the radiometric values of two images. Researchers have found that using different bands for image rationing can result in different performance (Chavez and MacKinnon, 1994; Howarth and Wickware, 1981; Jensen and Toll, 1982; Nelson, 1983). An advantage of the image rationing method is its robustness against radiometric differences due to illumination changes of the environment, which creates higher spectrum separability for CD. The change vector analysis (CVA) on multi-temporal CDs stacks the multi-spectral (Johnson and Kasischke, 1998) or multi-level feature differences (Tian *et al.*, 2013) in a vector form. This high-dimensional vector is analyzed in its directions and magnitude in the Euclidean space. The direction change can be used to easily infer the significance of bands that contribute to the change (Ye *et al.*, 2016). Each component of the vector can be also weighted according to its significance to the change description for improving detection accuracy. The background subtraction (BS) method assumes that the non-change areas usually have lower spectral/radiometric variances. Prior to the image difference, a low-pass filter is first applied to images to reduce spatial variances, and then the filtered image is subtracted from the original images (Singh, 1989). This method is easy to implement but has low accuracy, since important spatial features are also smoothed and subtracted.

Direct comparison on spectral bands may result in false alarms due to the ambiguity of the spectral information. Therefore, researchers are interested in transformation-based methods for image differencing (Collins and Woodcock, 1994; Fung and Ledrew, 1987; Parra *et al.*, 1996). The main advantage is that transformed methods can reduce the high-dimensional features space, and de-correlates inter-band relations (Lu *et al.*, 2004). PCA and TC transformation are orthogonal transformations that transform each component of the original signal into dimensions ordered via the variations. PCA is usually applied to the original image bands and TC is applied to the scene independent components such as brightness, greenness and wetness (Collins and Woodcock, 1994; Lu *et al.*, 2004; Munyati, 2004). Chen *et al.*, (2013) proposed using the gradient of the spectrum curve to compute the difference maps, which showed more accurate results than image differencing and the CVA method.

A major challenge for 2D CD in LTMR images is estimation of the signal-to-noise ratio of the images in order to eliminate the unwanted factors induced by the quality of the data. Moreover, even though LTMR images are supposed to have less perspective distortions, it has been reported that accurate pixel-wise co-registration of images is critical for obtaining reasonable detection accuracies (Lu *et al.*, 2004). Systematic reviews of 2D CD approaches can be found in Collins and Woodcock (1996), Hayes and Sader (2001), Lu *et al.* (2004), Radke *et al.* (2005) and Singh (1989). The sum of the squared spectral difference usually results in low separability of real change and unwanted changes, which sometimes are due to the correlation between the images bands. Transformation-based methods aim to remove the inter-band correlation for a better delineation of changes (Coppin and Bauer, 1996; Coppin *et al.*, 2001; Fung and Ledrew, 1987; Huang *et al.*, 2014; Lillesand *et al.*, 2004; Munyati, 2004). Nowadays 2D CD techniques on LTMR images have become pretty standard, and practical systems are available already in remote sensing software packages such as ERDAS (HEXAGON, 2014) , ENVI (EXELIS, 2014), and eCognition (Trimble, 2014).

### 8.3.2 Object-based approaches

The objects of interest in VHR images (e.g. IKONOS (1 m), GeoEye (0.5 m), Worldview1/2 (0.5 m), Worldview 3/4 (0.31 m), Pleiades (0.7 m)) may be composed of groups of pixels (Blaschke, 2010), while single pixels with different spectrums may belong to the same object. This leads to the well-known 'salt-and-pepper' noises when using pixel-based methods for CD or classification on these images (Yu *et al.*, 2006). Object-based methods are more robust towards this problem: it was noticed in (Fisher, 1997) that a single pixel might not be enough to support the image analysis comparing to pixel groups, and this idea is further supported by Bontemps *et al.* (2008), Hay and Castilla (2006, 2008), Hussain *et al.* (2013) and Longley (2002). The aim of object-based methods is to analyze images on a per-object basis, i.e. groups of pixels derived following certain criteria (such as textural, spectral/radiometric or semantic homogeneity). These objects can be labeled segments from vector data, or the image segments computed using image segmentation methods (Comaniciu and Meer, 2002). Object-based methods are normally more robust and require less computations (Chen *et al.*, 2012*a*; Qin and Fang, 2014).

In addition, the shape of the segments can be used to identify objects that are not easily differentiable using spectral information alone, e.g. building roofs and roads, open ground/plaza and narrow pedestrian paths.

Object-based methods nowadays have been incorporated into major remote sensing image processing packages (EXELIS, 2014; Trimble, 2014) for VHR data. There are several ways of defining an object (Tewkesbury *et al.*, 2015) for a bi-temporal dataset:

1. Image-object overlay: the segmentation is performed on one of the images, and the change analysis is performed within each segment (Comber *et al.*, 2004; Listner and Niemeyer, 2011).
2. Image-object comparison: the segmentations are performed on each of the images, change analyses are performed separately and then combined (Boldt *et al.*, 2012; Ehlers *et al.*, 2014).
3. Multi-temporal segmentation: the segmentation is performed on the entire time-series (Bontemps *et al.*, 2012; Teo and Shih, 2013). This includes the case of taking the intersections of individually segmented images (Tian *et al.*, 2013).

Increasingly, studies showed that object-based methods demonstrate improvements over pixel-based methods in both classification and CD. Yu *et al.* (2006) performed a detailed study by classifying vegetation types in airborne images using an object-based method and demonstrated good results towards 'salt and pepper' problems. Desclée *et al.* (2006) proposed an object-based method for change detection on forest inventory. The intention behind using the object-based method was to delineate the change more robustly using statistical testing from pixel values within each object, which reported more than 90% detection accuracy. Conchedda *et al.* (2008) adopted multi-resolution segmentation techniques on SPOT multi-spectral images to derive the objects for studying the change of mangrove species. Class-specific rules were derived to incorporate the spectral information in each level of the segmentation, where the mean and standard deviation of the spectral bands of transit classes were used for detecting changes, where high detection accuracy was reported even for small scatters of changes.

CD between 2D images and vector data (from geodatabase) naturally fits the scenario of geodatabase updating. Bouziani *et al.* (2010) proposed an object-based CD method to automatically detect the building changes between optical images and existing digital cartography data. They compared the extracted segments from the VHR images to the cartographic map based on a set of rules as expert knowledge for geodatabase updating, and reported 90% of the changes were detected. Similarly, Durieux *et al.* (2008) proposed an object-based building detection method from SPOT 5 data and compared it with the existing GIS database to study the urban sprawl phenomenon. Similar methods can be found in Niemeyer *et al.* (2008).

Although the advantages of object-based methods are widely recognized, a potential issue arises with the over- or under-segmentation problem. For example, for methods that compare the statistics within the objects, under-segmentation may saturate significant changes by incorporating too many unchanged pixels, while over-segmentation may result in changes that are purely induced by noise.

### 8.3.3 Classification-based approaches

Classification-based approaches generally refer to supervised or unsupervised methods (index-based) that tend to label the pixels/objects before or after the image-based CD. According to its position in the CD procedure, it can be further divided into pre-classification (PREC) and post-classification (POSTC) approaches. The PREC methods perform image classification independently for each date, and then take the labeled image as an input for type change identification (Mas, 1999). The POSTC methods first perform image comparison in pixel/object level to find initial change masks, and then discriminate between real changes and unwanted changes (e.g. seasonal varying vegetation) or false positives using classification methods (Pacifici *et al.*, 2007).

The basic argument of the PREC-based approaches is that the direct image comparison (difference of intensities and transformed features, etc.) is sometimes affected by the differences in physical conditions during the data capture. Understanding the scene with learning-based approaches may lead to improved CD accuracy (Al-Khudhairy *et al.*, 2005). Walter (2004) proposed a classification-based method to detect changes between the images and the GIS database. He first employed a maximum likelihood classifier (Foody *et al.*, 1992) to classify the multi-spectral image in object-level, where the training samples were derived from the GIS database, and then the labeled objects were used for updating the geo-database. A similar method can be found in Knudsen and Olsen (2003). Frauman and Wolff (2005) adopted a pixel-based classification method that performed independent classification for multi-spectral images of each date, and then derived the change maps by checking the consistency between their class labels. Conchedda *et al.* (2008) proposed to adopt classification-based method to assess the mangrove changes using SPOT XS data, and a high detection rate was reported. Similarly, Dronova *et al.* (2011) applied an object-based nearest neighborhood method to examine the ecosystem change of Poyang Lake in China, which showed its value in identifying type transition in ecosystem studies.

The POSTC-based CD methods aim to separate real changes from false alarms or unwanted changes using learning-based strategy. Chaabouni-Chouayakh and Reinartz (2011) applied support vector machine (SVM) learning to separate the changes in trees and buildings from the initial change mask. Pacifici *et al.* (2007) proposed a two-step classification method for CD, which combined both PREC and POSTC strategy. They first performed a supervised classification using neural-network on the images from different dates independently, and then generated the change maps by comparing their class labels. The two results were then fused by intersecting both change maps.

Classification-based methods can by-pass the uncertainties induced by spectral comparison of images to a certain extent, since the semantics of each pixel/object are learned from the samples in the current images. However, CD results are closely related to the classification results. They might vary with the choice of samples, and with the granularity of the image segments, which can be sensitive to the parameters and image quality.

## 8.4 Change detection with 3D data

3D data are becoming increasingly available at reduced cost (Gehrke *et al.*, 2010). Low-cost LiDAR data, more automated acquisition and processing pipeline for airborne, UAV-borne (unmanned aerial vehicles) and space-borne data (Choudhary *et al.*, 2010; Hirschmüller, 2008; Nex and Remondino, 2014; Qin, 2016) are raising interest in change detection (CD) applications in 3D infrastructure monitoring (Grigillo *et al.*, 2011; Nebiker *et al.*, 2014), volumetric landslide monitoring (Martha *et al.*, 2010), disaster management (Menderes *et al.*, 2015), glacial monitoring (Haala and Rothermel, 2012b), and construction monitoring (Siebert and Teizer, 2014). The major advantages of using 3D data over 2D for CD are threefold:

1. *Insensitive to illumination differences:* The 3D data present the spatial measurements of the objects, therefore comparison of the geometry of bi-temporal data is irrespective of illumination conditions.
2. *Insensitive to perspective distortions in 2D CD:* Comparison of the geometry can be performed in a true three-dimensional space, or any projected space (subspace of 3D space). Oriented images can be easily corrected based on the available 3D data.
3. *Volumetric information:* 3D CD provides volumetric changes that can facilitate more applications, such as volumetric forest loss, accurate construction progress monitoring, etc.

The advantage of 3D data for CD has been known for a long time (Murakami *et al.*, 1999), while 3D CD that research has only recently been published, largely driven by the development of automated image-based 3D reconstruction (Remondino *et al.*, 2014). Comparison of bi-temporal data in the height or depth direction obviously leads to more robust results. It is easy to extend some of the 2D CD methods to 3D CD. For instance, pixel/object/classification-based methods can all apply to 3D data in a project space (raster forms), such as raster DSM, depth maps. However, the presence of information from an extra single dimension does not necessarily define 3D CD as a simple extension of 2D CD, given more complicated issue it brings in:

1. *Uncertainties of 3D data:* 3D data generation (e.g. from images) brings different types of uncertainties, associated with the image quality, convergence angle, DIM algorithms and object scale. For example, the image matching may fail on thin and tall objects or large texture-less areas. Uncertainties of point clouds generated using different dense matching methods may have different and non-uniform distributions.
2. *Multi-modal data fusion:* Geometric data present a different modality from the image data. Fusion of both forms of data requires special considerations of different types of data uncertainties, feature extraction and multi-source weighting (Tian *et al.*, 2013).
3. *CD under a true 3D environment:* 2D CD methods can be extended straightforwardly to process 2.5D data such as bi-temporal raster DSMs or depth maps. However, in a truly 3D environment (including height and façade reliefs,

sometimes non-convex and non-watertight) – e.g. 3D data generated from oblique imagery, LiDAR point clouds from mobile/terrestrial laser scanning – 2D CD methods can hardly solve the problem. Moreover, the presence of occlusions, disturbances of unwanted objects, incomplete data, and 3D feature extraction in such scenarios, requires new CD techniques and methods.

Technically speaking, although this additional dimension brings a more complex scenario for CD, there are two fundamental aspects that can simply encapsulate the particularities of 3D CD techniques: (1) geometric comparison; and (2) geometric-spectral analysis. Geometric comparison refers to methods that measure geometric differences of 3D datasets for change determination, while geometric-spectral analysis considers both the geometry and spectrum (colour) fusion for CD. These two aspects can also be incorporated into a single 3D CD framework. In addition, 2D image-based (raster-based) techniques can be a valuable reference for processing 3D data in a projected space (2.5D data, e.g. height, depth maps). In this section, we largely follow our previous review in Qin *et al.* (2016a) and focus on techniques that perform geometric differencing under difference scenarios, as well as methods for fusing both geometric and spectral data for CD.

### 8.4.1 Geometric comparison

The geometric comparison varies greatly with the viewing scenario (oblique-view, top-view) and data format (DSM, point clouds, stereo images, etc.). It refers to a 2.5D comparison such as height/depth difference (shown in Figure 8.1a), or a full 3D comparison through a Euclidean distance measure (shown in Figure 8.1b). Moreover, image sets taken from different perspectives implicitly contain 3D geometric information (3D-IMP), and the geometric difference in such data requires image comparison through projection (projection-based method) (an example is shown in Figure 8.1c) or multi-ray consistency evaluation. There is not a best out of the three: each can be applied to an appropriate context depending on the viewing scenario and data formats.

*Height differencing:* Height differencing is the most basic and the simplest type of geometric comparison, usually applied to bi-temporal raster-DSM or depth maps. This essentially treats the height differencing as a raster comparison problem, where pixel/object-based differencing methods can be directly used. Per-pixel height differencing has been widely used in applications such as tree growth monitoring (Gong

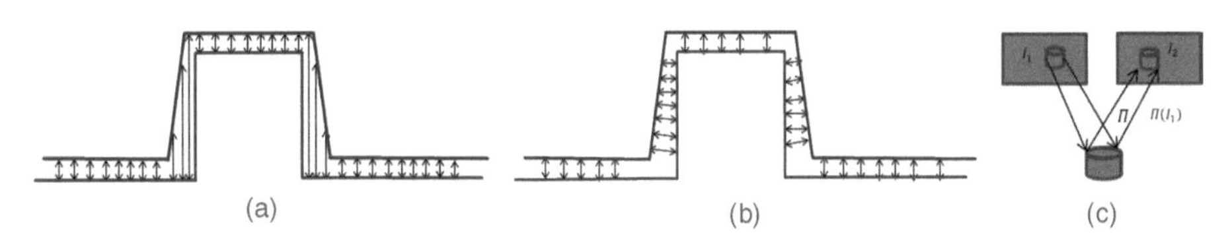


Figure 8.1 Different geometric comparison methods. (a) Height difference: distances are computed vertically. (b) Euclidean distances: distances are computed in the surface normal direction. (c) Projection-based inter-correlation method: the geometric difference is computed by projecting image $I_1$ on to the object, and then back projecting to image $I_2$ as $\Pi(I_1)$; the differences are given by measuring the differences between $\Pi(I_1)$ and . $I_2$

*et al.*, 2000; Stepper *et al.*, 2015; Waser *et al.*, 2007, 2008), earthquakes and damage assessment (Strong, 1974; Turker and Cetinkaya, 2005), etc. It has also been applied to urban areas. Sasagawa *et al.* (2013) applied height differencing in an urban area using DSMs generated from ALOS (Advanced Land Observation Satellite) triplets to indicate changes in individual buildings. Due to the presence of noise in the DSM or depth maps, window-based or object-based methods can be used to average the height differences (Tian *et al.*, 2010, 2013). These objects can be derived from available ortho-photos associated with the DSM (Tian *et al.*, 2013), as the images normally contain richer information of the object boundary. Moreover, refinements can be performed using additional features such as geometric primitives, textural/spectral features, or external data sources such as from a GIS (Geographical Information System) database (Dini *et al.*, 2012).

The height threshold, similar to thresholds of image differencing, is one of the most important parameters to determine the final changes. The selection of the threshold is correlated with the accuracy of the DSM. One means of threshold determination is to use a priori information such as the pre-assessment of the DSM quality and empirical choices, or trial-and-error tests (Lu *et al.*, 2004; Murakami *et al.*, 1999). Another way is to estimate the threshold based on statistical measures, such as from the histogram of the height residuals (Chaabouni-Chouayakh and Reinartz, 2011). The height threshold reflects the actual metric of objects; it is, however, very often data-dependent, which is again due to the different level uncertainty of the bi-temporal 3D data. To avoid single threshold truncation, multiple thresholds can be also used to indicate different levels of confidence (Qin *et al.*, 2015). Regions with a very high confidence of being changed can be used directly as the CD output, while uncertain ones could be sent for the operator's decision.

*3D Euclidean distances*: The Euclidean distance between two surfaces takes the three degrees of freedom (of the 3D geometry) into account by computing the distance along the normal direction, which is theoretically more rigorous. The difference between the Euclidean and height distances can be easily understood in Figure 8.1a–b. The Euclidean distance accounts better for the misregistration/DSM errors in the building boundaries by considering the normal direction. Techniques in this category are generally developed in the domain of surface co-registration and CD, where, in surface co-registration, changes are regarded as outliers. An example of such technique is the least squares 3D surface matching proposed by Gruen and Akca (2005), which performs 3D data co-registration using Euclidean distance measures. It was later applied by Waser *et al.* (2008) to estimate the forest volume dynamics between two image-derived DSMs. Under the context of 3D model quality control, Akca *et al.* (2010) adopted the LS3D method to detect the 3D geometric modelling error against the LiDAR measurements, where Euclidean differences were estimated by minimizing the Euclidean distances between the 3D data. For similar methods for CD, readers may refer to the global methods that minimize point-to-surface or surface-to-surface distances, where the outliers of the co-registration are detected as changes (Habib *et al.*, 2005; Karras and Petsa, 1993; Maas, 2000; Mitchell and Chadwick, 1999; Pilgrim, 1996; Rosenholm and Torlegard, 1988).

Occlusions and incompleteness of 3D data (from LiDAR or images) present much more complex scenarios than remote sensing top-view data. Akca (2007) showed various successful CD examples using the Euclidean distance measure in close-range applications under the context of deformation analysis and quality control (Akca *et al.*, 2010). Other derivative measures based on Euclidean distance can be also used for CD: Girardeau-Montaut *et al.* (2005) applied an octree structure to divide the 3D spaces, and the Hausdorff measure was employed to compute the distance between the different spaces. Similarly, Kang and Lu (2011) adopted the Hausdorff distance to detect the difference between LiDAR scanning data and a reference 3D model.

Sometimes the Euclidean distance between extracted geometric features appears to be more robust than per-point Euclidean distances: Eden and Cooper (2008) measured the differences of 3D lines across two multi-view image sets, which significantly reduced the noise and disturbances. Utilizing the same concept, Champion *et al.* (2010) extracted 3D lines from stereo images to verify the existence of buildings by comparing them to the GIS database. Nevertheless, failing to detect such features may omit some important changes. Therefore, feature-based methods should only be applied in the context where the object of interest can be represented by certain features.

*Projection-based geometric differences*: It is not always the best recommendation to generate 3D point clouds/DSM from stereo/multi-stereo images before the CD step. Stereo images taken under suboptimal conditions (poor intersection angle and high environmental illumination differences) may produce unreliable 3D information. If relatively reliable DSM or point clouds are available at one date and images are oriented with respect to the 3D data, the projection-based geometric difference can be used to assess the geometric consistency without explicitly generating 3D data from the images. It correlates one image of the stereo pair with the other image, using the DSM or point cloud, and compares their radiometric/spectral differences (shown in Figure 8.1c). In principle, these two correlated images should be the same if the stereo pair is consistent with the DSM/point clouds. Qin (2014a) applied inter-correlation in the process of 3D model updating, where two stereo satellite images were correlated using 3D polygonal models, and the correlated image patches were evaluated using the energy produced by the SGM (semi-global matching) algorithm (Hirschmüller, 2008). In Knudsen and Olsen (2003), 3D models were projected onto 2D photos, followed by supervised classification for CD.

This technique is particularly effective for oblique-view images and point clouds/3D models, as a direct comparison using point clouds generated from such images via DIM usually produces many artifacts. Taneja *et al.* (2011) applied inter-correlation of a stereo pair to an image-derived surface model, and the differences in colour were used as evidence of change. Qin and Gruen (2014) extended inter-correlation to a multi-stereo case in order to determine view-based change evidence by comparing a strip of images with mobile LiDAR point clouds.

Another streamlining of the projection-based method divides the 3D spaces into voxel/object representations, where photo-consistencies of multi-view images projected to the cells are evaluated statistically. Voxels with significant colour variations will be spotted as changes; examples are Crispell *et al.* (2012), LeCun *et al.* (2015), Pollard and Mundy (2007) and Schindler and Dellaert (2010). Such colour-consistency checking implicitly applies a multi-ray point matching strategy, where false positives might be present in occluded areas and false negatives might occur in non-texture areas. After probability assignment, Markov interfering processes (Blake *et al.*, 2011) are applied, very often to reduce noise effects.

The projection-based method is an effective strategy for providing raw change evidence when the 3D scene is rather complex, e.g. oblique data or street-view data. It can be seen as an inverse operation to matching, while this again, still depends on the quality of the available 3D data and it may not be able to handle areas with insignificant texture features.

## 8.4.2 Geometric-spectral analysis

Usually, geometric 3D data come along with spectral/intensity information, e.g. orthophotos associated with the DSM, textures associated with the 3D models. It is straightforward to understand that additional channels of information may lead to enhanced CD results, as the geometric and spectral information could be beneficial to each other, while on the other hand, it faces the risk of error propagation from both sources. The main challenge, therefore, in this case lies in the information fusion for optimal signal-noise ratio (utilize useful information of both sources while suppressing their errors). In general, there are three ways to integrate the geometric and spectral features into a 3D CD process:

1. Post-refinement.
2. Direct feature fusion.
3. Classification-based method.

Post-refinement refers to methods that perform refinement on initial change evidence from geometry and/or spectrum comparison (Chaabouni-Chouayakh *et al.*, 2011). Direct feature fusion jointly fuses geometric and spectral information (or their transformed features) in the feature level, and the fused features are used to determine the presence of change. The third approach, the classification-based method, is very similar to the classification-based method in 2D CD, which first classifies both datasets or detects the objects of interest using 3D object detection methods (Qin *et al.*, 2015), and then compares the resulting labels of the two datasets.

*Post-refinement*: False positives/negatives from geometric comparison occur due to artifacts of the DSM/point clouds, or incomplete 3D models. Sometimes the errors follow certain patterns, e.g. artifacts of building changes often occur at object boundaries or in vegetation classes (due to failures of DIM methods for complex geometry). These can be well addressed if additional information/features can be extracted from the geometric or spectral data. If near-infrared data is available, NDVI (normalized difference vegetation index) can be used to eliminate disturbances from the seasonal varying vegetation. Such an attempt was performed through manual interpretations (Sasagawa *et al.*, 2008), where the radiometric difference of the images was used as a 'double-check' for DSM subtraction results. To automate the process, change 'candidates' can be further classified by using spectral and textural information of the original images (Fan *et al.*, 1999; Liu *et al.*, 2003; Pang *et al.*, 2014).

Due to the presence of noise in DSM subtraction, some specific noise-removal approaches – for instance, morphological filtering – can be used to improve the initial change masks (Chaabouni-Chouayakh *et al.*, 2010; Choi *et al.*, 2009; Zhu *et al.*, 2008). External data such as map boundaries (James *et al.*, 2012) can be used to constrain the CD masks for certain types of objects. Based on the assumption that the change maps are globally smooth, Guerin *et al.* (2014) applied a global optimization procedure that employs this spatial context using a generalized dynamic programming to eliminate potential inaccuracies resulting from DSM subtractions. Markov random field (MRF) approaches as used in 2D CD approaches (Diakité *et al.*, 2014; Huang, 2013) were also developed under the 3D context (Pollard and Mundy, 2007; Qin and Gruen, 2014; Tornabene *et al.*, 1983). Using UAV (unmanned aerial vehicle) images for CD, Qin (2014b) hierarchically refined the initial change masks using various multi-level segmentations from ortho-photos and DSM. This work refined the mask using the spatial consistencies of these segments, and reported that the method can monitor even sub-building sized urban objects (such as vehicles).

Post-refinement approaches employ a hierarchical processing structure, where initial change evidence from geometric comparison, is refined based on available geometric and spectral features. Parameters relevant to the geometry are often easy to understand and straightforward to tune. Such 'step-wise' methods are flexible and can be decomposed or re-composed according to different CD applications. However, the initial CD result depends solely on the geometric comparison, and missing changes in the initial step cannot be recovered in the subsequent refinement.

*Direct feature fusion:* Instead of processing geometric and spectral information hierarchically, direct feature fusion simultaneously considers all channels of information. Such feature fusion can be performed in either the feature level or decision level. Although existing work in 'direct feature fusion' mainly considers the fusion of multi-source spectral data (Hartman, 2008; Longbotham *et al.*, 2012) for CD, there is still some work that fuses both geometric and spectral information directly. Tian *et al.* (2013) directly fused the height and radiometric differences of Cartosat-1 datasets (only panchromatic images are available) under a change vector analysis (CVA) framework. Their subsequent work (Tian *et al.*, 2014) adopted a Kernel Minimum Noise Fraction (KMNF) to minimize the noise statistically presented in both the height and radiometric difference for fusion, and an Iterated Canonical Discriminant Analysis (ICDA) for generating the final CD results. They reported that notable improvements were obtained in their forest CD applications using Cartosat-1 against methods that included simple DSM differencing, CVA fusion, and other traditional classification-based methods (Remondino and El-Hakim, 2006; Wang, 2005; Young *et al.*, 2010). Under a 3D model updating process, Qin (2014a), fused multiple change evidences resulting from DSM and spectral features via unsupervised self-organizing maps (SOM) (Kohonen, 1982; Moosavi and Qin, 2012). It has been showed that the a priori information (the quality of the change evidence) can be used to weight individual change indicators for a better change determination.

Supervised approaches that use a feature vector stacking different geometric-spectral features also fall into the 'direct feature fusion' category (Chehata *et al.*, 2009; Chen *et al.*, 2012*b*; Nemmour and Chibani, 2006; Pacifici *et al.*, 2007). These methods consider both the geometric and spectral data as pure information sources. Other, different kinds of information can be easily incorporated into the classifier without additional re-design of the algorithm. However, it is critical to determine the individual contribution of each source when using linear fusion models. Classifier-based models may be able to learn the weights of information sources, such as random forests

(Breiman, 2001) and neural network (Foody, 1996), while this requires accurate training samples. For unsupervised fusion models (e.g. CVA), an equal contribution may not render the best results. Therefore a priori information or a trial-and-error test may be needed to obtain an optimal parameter configuration.

*Classification-based method:* Accuracy, texture and spectral differences in temporal 3D data may bring errors in direct geometric and/or spectral comparison processes. The classification-based methods propose detecting objects of interest or performing land-cover classification first, and then comparing the resulting labels (classes), which avoids direct comparison of the spectral and height information. These methods share the same idea as the classification-based approaches introduced in Section 8.3.3; the core difference is that the additional geometric information may represents a different modality data, which might have the potential to enhance both the classification and object detection accuracy for CD. A number of studies (Huang *et al.*, 2011; Mayer, 1999; Sohn and Dowman, 2007; Zhang *et al.*, 2015) have proven that height information can increase the accuracy of land-cover classification to a notable level. The DSMs from LiDAR or stereo images can essentially be seen as an additional channel of information, which is equally free to be applied into popular classifiers. Researchers have investigated such a strategy via a number of classification approaches, such as in ISODATA (Olsen, 2004), maximum likelihood (Walter, 2004), decision tree (Matikainen *et al.*, 2010), rule-based method (Champion, 2007; Olsen and Knudsen, 2005) and decision-fusion method (Nebiker *et al.*, 2014; Rottensteiner *et al.*, 2007).

In an urban environment, building change detection for geodatabase updating is the most relevant application. 'Building detection + Change detection' is a popular strategy for detecting changes in buildings. Under this framework, Qin *et al.* (2015) integrated height information with a supervised framework to detect buildings using scanned aerial survey photos. Detected building objects from each date were then compared by considering both the height and texture dissimilarities. The height information was implemented in different levels of processing, including image segmentation, classification and CD, which was proven to be particularly effective for rebuilt buildings as it evaluates each building object using various features such as height, texture, as well as shapes. Moreover, existing GIS data can be used as training data (Champion *et al.*, 2009; Matikainen *et al.*, 2010; Walter, 2004) to assist building detection and subsequently for updating. They can either be used directly as training samples (Walter, 2004), or modified using some other cues based on geometric and spectral features (Champion *et al.*, 2009; Qin *et al.*, 2016b). The classification method, including its 2D counterpart, is so far regarded as the popular method, since it transforms the direct geometric/spectral comparison to label changes, which offer information regarding type changes. However, in most cases, the CD results of this method depend highly on the classification/object detection results, which subsequently requires careful sample collection and feature design.

## 8.5 Summary

The creation of a geodatabase (2D vector map, 3D city models) is nowadays regarded as a necessary step for initiating digital infrastructures for smart city management. While the techniques for creating engineering-grade geodatabases remain a time-consuming process, the need to update geodatabases to keep pace with the reality is at stake. This calls for efficient change detection (CD) solutions that are capable of identifying the changed area in a certain period, in order to rebuild and update the 2D/3D vector data. In the meantime, temporal change itself acts as a very valuable means for assessing the dynamics of land processes and urban development. This chapter has provided an overview of state-of-the-art change detection methods, structured intuitively as both 2D and 3D data change detection. But 2D CD and 3D CD approaches are not necessarily separate from each other, as many methods for processing 2.5D data are indeed very similar to 2D CD methods.

CD is a disparate subject that involves many complicated issues due to the levels of noise, accuracy, acquisition conditions, data format and availability of auxiliary data. Therefore, it is rather complex to conclude that any one approach is superior to others. That 3D CD is more robust than 2D CD seems to be an obvious and consistent conclusion, but existing work still lacks a good comparative study that analyzes comprehensively the performance of spectral and geometric data for CD under different conditions and scenarios, including comparison using similar data under similar application contexts. However, so far, there is general agreement on the prevalent methods and their performance,, including pixel-based, object-based and classification-based methods. It seems that this ad-hoc CD method for applications is still the one mostly used, particularly for the task of updating geodatabases, where the availability of source data (images and LiDAR point clouds) and the format of vector data may vary greatly. Where semantic information on the geodatabase is available, it is recommended that information such as building block number, streets, number of floors, types of objects, and types of buildings, is incorporated into the CD algorithm to render more robust/engineered solutions.

As we look at the CD algorithms themselves, '*comparison* of data for changes on *object of interest*' is key. These seemingly independent keywords are intimately connected to three fundamental issues:

1. Advanced spectrum correction and comparison methods.
2. Advanced 3D data generation algorithms that provide more accurate 3D measurements for accurate geometric comparison,
3. High-level feature extraction and machine-learning methods for detecting the object of interest.

Therefore, in addition to efforts to reduce noise and unwanted changes, advanced CD algorithms must rely upon future endeavor, with regard to these three aspects, to generate more integrated and robust methods for geodatabase updating.

## Acknowledgement

Parts of this chapter are taken from the author's PhD thesis, '3D change detection in an urban environment with multi-temporal data' and a review paper, '3D change detection: Approaches and applications'.

## References

Akca, D. (2007) Least squares 3D surface matching. PhD thesis, Swiss Federal Institute of Technology, ETH, Zürich.

Akca, D., M. Freeman, I. Sargent and A. Gruen (2010) Quality assessment of 3D building data. *The Photogrammetric Record* **25**(132), 339–355.

Al-Khudhairy, D., I. Caravaggi and S. Giada (2005) Structural damage assessments from Ikonos data using change detection, object-oriented segmentation, and classification techniques. *Photogrammetric Engineering and Remote Sensing* **71**(7), 825–837.

Baltsavias, E. P. (1991) Multiphoto geometrically constrained matching. Ph.D thesis, Instittute of Geodesy and Photogrammetry, Swiss Federal Institute of Technology, ETH, Zürich.

Berk, A., G. P. Anderson, L. S. Bernstein, P. K. Acharya, H. Dothe, M. W. Matthew, S. M. Adler-Golden, J. H. Chetwynd Jr, S. C. Richtsmeier and B. Pukall (1999) MODTRAN 4 radiative transfer modelling for atmospheric correction. In: *Proceedings of SPIE – The International Society for Optical Engineering*, vol. 3756, 348–353.

Besl, P. and N. Mckay (1992) A method for registration of 3-D shapes. *IEEE Transactions on Pattern Analysis and Machine Intelligence* **14**(2), 239–256.

Bioucas-Dias, J. M., A. Plaza, N. Dobigeon, M. Parente, Q. Du, P. Gader and J. Chanussot (2012) Hyperspectral unmixing overview: Geometrical, statistical, and sparse regression-based approaches. *IEEE Journal of Selected Topics in Applied Earth Observations and Remote Sensing* **5**(2), 354–379.

Blake, A., P. Kohli and C. Rother (2011) *Markov Random Fields for Vision and Image Processing*. MIT Press, pp. 472.

Blaschke, T. (2010) Object based image analysis for remote sensing. *ISPRS Journal of Photogrammetry and Remote Sensing* **65**(1), 2–16.

Boldt, M., A. Thiele and K. Schulz (2012) Object-based urban change detection analyzing high resolution optical satellite images. *Proc. SPIE Remote Sens* 85380E-1.

Bontemps, S., P. Bogaert, N. Titeux and P. Defourny (2008) An object-based change detection method accounting for temporal dependences in time series with medium to coarse spatial resolution. *Remote Sensing of Environment* **112**(6), 3181–3191.

Bontemps, S., A. Langner and P. Defourny (2012) Monitoring forest changes in Borneo on a yearly basis by an object-based change detection algorithm using SPOT-VEGETATION time series. *International Journal of Remote Sensing* **33**(15), 4673–4699.

Bouziani, M., K. Goïta and D.-C. He (2010) Automatic change detection of buildings in urban environments from very high spatial resolution images using existing geodatabase and prior knowledge. *ISPRS Journal of Photogrammetry and Remote Sensing* **65**(1), 143–153.

Breiman, L. (2001) Random forests. *Machine Learning* **45**(1), 5–32.

Chaabouni-Chouayakh, H., P. d'Angelo, T. Krauss and P. Reinartz (2011) Automatic urban area monitoring using digital surface models and shape features. In: *Joint Urban Remote Sensing Event (JURSE)*, 2011, 85–88.

Chaabouni-Chouayakh, H., T. Krauss, P. d'Angelo and P. Reinartz (2010) 3D change detection inside urban areas using different digital surface models. *International Archives of Photogrammetry, Remote Sensing and Spatial Information Sciences* **38**, 86–91.

Chaabouni-Chouayakh, H. and P. Reinartz (2011) Towards automatic 3D change detection inside urban areas by combining height and shape information. *Photogrammetrie-Fernerkundung-Geoinformation* (4), 205–217

Champion, N. (2007) 2D building change detection from high resolution aerial images and correlation digital surface models. *International Archives of Photogrammetry, Remote Sensing and Spatial Information Sciences* **36**(3/W49A), 197–202.

Champion, N., D. Boldo, M. Pierrot-Deseilligny and G. Stamon (2010) 2D building change detection from high resolution satelliteimagery: A two-step hierarchical method based on 3D invariant primitives. *Pattern Recognition Letters* **31**(10), 1138–1147.

Champion, N., F. Rottensteiner, L. Matikainen, X. Liang, J. Hyyppä and B. Olsen (2009) A test of automatic building change detection approaches. *International Achieves of Photogrammetry, Remote Sensing and Spatial Information Sciences* **38**(Part 3/W4), 145–150.

Chavez, P. S. and D. J. MacKinnon (1994) Automatic detection of vegetation changes in the southwestern United States using remotely sensed images. *Photogrammetric Engineering and Remote Sensing* **60**(5), 571–583.

Chehata, N., L. Guo and C. Mallet (2009) Airborne lidar feature selection for urban classification using random forests. *International Archives of Photogrammetry, Remote Sensing and Spatial Information Sciences* **38**(Part 3), W8.

Chen, G., G. J. Hay, L. M. Carvalho and M. A. Wulder (2012a) Object-based change detection. *International Journal of Remote Sensing* **33**(14), 4434–4457.

Chen, G., K. Zhao and R. Powers (2014) Assessment of the image misregistration effects on object-based change detection. *ISPRS Journal of Photogrammetry and Remote Sensing* **87** (2014), 19–27.

Chen, J., M. Lu, X. Chen, J. Chen and L. Chen (2013) A spectral gradient difference based approach for land cover change detection. *ISPRS Journal of Photogrammetry and Remote Sensing* **85**(2013), 1–12.http://dx.doi.org/10.1016/j.isprsjprs.2013.07.009.

Chen, L., S. Zhao, W. Han and Y. Li (2012b) Building detection in an urban area using lidar data and QuickBird imagery. *International Journal of Remote Sensing* **33**(16), 5135–5148

Choi, K., I. Lee and S. Kim (2009) A feature based approach to automatic change detection from Lidar data in urban areas. *International Archives of Photogrammetry, Remote Sensing and Spatial Information Sciences* **38**(Part 3/W8), 259–264.

Choudhary, S., S. Gupta and P. Narayanan (2010) Practical time bundle adjustment for 3D reconstruction on the GPU. In: *Trends and Topics in Computer Vision. ECCV 2010.* Lecture Notes in Computer Science, vol. 6554. Berlin, Heidelberg: Springer, 423–435.

Collins, J. B. and C. E. Woodcock (1994) Change detection using the Gramm-Schmidt transformation applied to mapping forest mortality. *Remote Sensing of Environment* **50**(3), 267–279.

Collins, J. B. and C. E. Woodcock (1996) An assessment of several linear change detection techniques for mapping forest mortality using multitemporal Landsat TM data. *Remote Sensing of Environment* **56**(1), 66–77.

Comaniciu, D. and P. Meer (2002) Mean shift: A robust approach toward feature space analysis. *IEEE Transactions on Pattern Analysis and Machine Intelligence* **24**(5), 603–619.

Comber, A., P. Fisher and R. Wadsworth (2004) Assessment of a semantic statistical approach to detecting land cover change using inconsistent data sets. *Photogrammetric Engineering & Remote Sensing* **70**(8), 931–938.

Conchedda, G., L. Durieux and P. Mayaux (2008) An object-based method for mapping and change analysis in mangrove ecosystems. *ISPRS Journal of Photogrammetry and Remote Sensing* **63**(5), 578–589.

Coppin, P. R. and M. E. Bauer (1996) Digital change detection in forest ecosystems with remote sensing imagery. *Remote Sensing Reviews* **13**(3–4), 207–234.

Coppin, P., I. Jonckheere, K. Nackaerts, B. Muys and E. Lambin (2004) Digital change detection methods in ecosystem monitoring: A review. *International Journal of Remote Sensing* **25**(9), 1565–1596.

Coppin, P., K. Nackaerts, L. Queen and K. Brewer (2001) Operational monitoring of green biomass change for forest management. *Photogrammetric Engineering and Remote Sensing* **67**(5), 603–612.

Crispell, D., J. Mundy and G. Taubin (2012) A variable-resolution probabilistic three-dimensional model for change detection. *IEEE Transactions on Geoscience and Remote Sensing*, **50**(2), 489–500.

d'Angelo, P. and P. Reinartz (2011) Semiglobal matching results on the ISPRS stereo matching benchmark. *Int. Arch. Photogramm. Remote Sens. Spatial Inf. Sci., XXXVIII-4/W19*, 79–84, https://doi.org/10.5194/isprsarchives-XXXVIII-4-W19-79-2011.

Desclée, B., P. Bogaert and P. Defourny (2006) Forest change detection by statistical object-based method. *Remote Sensing of Environment* **102**(1), 1–11.

Diakité, A. A., G. Damiand and D. Van Maercke (2014) Topological reconstruction of complex 3D buildings and automatic extraction of levels of detail. In: *Proceedings of the 2nd Eurographics Workshop on Urban Data Modelling and Visualisation*, 25–30.

Dini, G., K. Jacobsen, F. Rottensteiner, M. Al Rajhi and C. Heipke (2012) 3D building change detection using high resolution stereo images and a GIS database. *ISPRS International Archives of the Photogrammetry, Remote Sensing and Spatial Information Sciences* **1**, 299–304.

Dronova, I., P. Gong and L. Wang (2011) Object-based analysis and change detection of major wetland cover types and their classification uncertainty during the low water period at Poyang Lake, China. *Remote Sensing of Environment* **115**(12), 3220–3236.

Durieux, L., E. Lagabrielle and A. Nelson (2008) A method for monitoring building construction in urban sprawl areas using object-based analysis of Spot 5 images and existing GIS data. *ISPRS Journal of Photogrammetry and Remote Sensing* **63**(4), 399–408.

Eden, I. and D. B. Cooper (2008) Using 3D line segments for robust and efficient change detection from multiple noisy images. In: *10th European Conference on Computer Vision, Marseille, France, 12–18, October*, 172–185.

Ehlers, M., N. Sofina, Y. Filippovska and M. Kada (2014) Automated techniques for change detection using combined edge segment texture analysis, GIS, and 3D information. In *Global Urban Monitoring and Assessment through Earth Observation*, Qihao Weng (ed.), CRC Press, 325–351.

El Hajj, M., A. Bégué, B. Lafrance, O. Hagolle, G. Dedieu and M. Rumeau (2008) Relative radiometric normalization and atmospheric correction of a SPOT 5 time series. *Sensors* **8**(4), 2774–2791.

EXELIS (2014) ENVI, http://www.exelisvis.com/ProductsServices/ENVIProducts/ENVI.aspx. (last accessed: 26 November 2014).

Fan, H., J. Zhang, Z. Zhang and Z. Liu (1999) House change detection based on DSM of aerial image in urban area. *Geo-spatial Information Science* **2**(1), 68–72.

Fischler, M. A. and R. C. Bolles (1981) Random sample consensus: A paradigm for model fitting with applications to image analysis and automated cartography. *Communications of the ACM* **24**(6), 381–395.

Fisher, P. (1997) The pixel: A snare and a delusion. *International Journal of Remote Sensing* **18**(3), 679–685.

Foody, G. M. (1996) Relating the land-cover composition of mixed pixels to artificial neural network classification output. *Photogrammetric Engineering and Remote Sensing* **62**(5), 491–498.

Foody, G. M., N. Campbell, N. Trodd and T. Wood (1992) Derivation and applications of probabilistic measures of class membership from the maximum-likelihood classification. *Photogrammetric Engineering and Remote Sensing* **58**(9), 1335–1341.

Fraser, C. S. and H. B. Hanley (2003) Bias compensation in rational functions for IKONOS satellite imagery. *Photogrammetric Engineering and Remote Sensing* **69**(1), 53–58.

Frauman, E. and E. Wolff (2005) Change detection in urban areas using very high spatial resolution satellite images: A case study in Brussels. In: *Proceedings of the 25th Annual Symposium of Remote Sensing Laboratories, Porto, Portugal, June 6–11*, EARSeL: EUROPEAN ASSOCIATION OF REMOTE SENSING LABORATORIES, 557–566.

Fung, T. (1990) An assessment of TM imagery for land-cover change detection. *IEEE Transactions on Geoscience and Remote Sensing* **28**(4), 681–684.

Fung, T. and E. Ledrew (1987) Application of principal components analysis to change detection. *Photogrammetric Engineering and Remote Sensing* **53**(12), 1649–1658.

Furukawa, Y. and J. Ponce (2010) Accurate, dense, and robust multiview stereopsis. *IEEE Transactions on Pattern Analysis and Machine Intelligence*, **32**(8), 1362–1376.

Gehrke, S., K. Morin, M. Downey, N. Boehrer and T. Fuchs (2010) Semi-global matching: An alternative to LIDAR for DSM generation? *International Archives of the Photogrammetry, Remote Sensing and Spatial Information Sciences*, Calgary, AB, **38**(B1) 6.

Girardeau-Montaut, D., M. Rouxa, R. Marcb and G. Thibaultb (2005) Change detection on point cloud data acquired with a ground laser scanner. *The International Archives of Photogrammetry, Remote Sensing and Spatial Information Sciences* **36**(Part 3/W19), 30–35.

Gong, P., G. S. Biging and R. Standiford (2000) Technical note: Use of digital surface model for hardwood rangeland monitoring. *Journal of Range Management* **53**(6), 622–626.

Gordon, H. R. (1997) Atmospheric correction of ocean colour imagery in the Earth Observing System era. *Journal of Geophysical Research: Atmospheres* **102**(D14), 17081–17106.

Grigillo, D., M. Kosmatin Fras and D. Petrovič (2011) Automatic extraction and building change detection from digital surface model and multispectral orthophoto. *Geodetski Vestnik* **55**(1), 28–45.

Gruen, A. and D. Akca (2005) Least squares 3D surface and curve matching. *ISPRS Journal of Photogrammetry and Remote Sensing* **59**(3), 151–174.

Guerin, C., R. Binet and M. Pierrot-Deseilligny (2014) Automatic detection of elevation changes by differential DSM analysis: Application to urban areas. *IEEE Journal of Selected Topics in Applied Earth Observations and Remote Sensing* **7**(10), 4020–4037.

Haala, N. and M. Rothermel (2012a) Dense multiple stereo matching of highly overlapping UAV imagery. *ISPRS International Archives of the Photogrammetry, Remote Sensing and Spatial Information Sciences* **39**(B1), 387–392.

Haala, N. and M. Rothermel (2012b) Dense multiple stereo matching of highly overlapping UAV imagery. *ISPRS International Archives of the Photogrammetry, Remote Sensing and Spatial Information Sciences* **39**(B1).

Habib, A., M. Ghanma, M. Morgan and R. Al-Ruzouq (2005) Photogrammetric and LiDAR data registration using linear features. *Photogrammetric Engineering and Remote Sensing* **71**(6), 699–707.

Hartley, R. and A. Zisserman (2004) *Multiple view Geometry in Computer Vision*. Cambridge: Cambridge University Press, 655pp.

Hartman, E. (2008) A promising oil alternative: Algae energy. *The Washington Post* 6 12.

Haverkamp, D. and R. Poulsen (2003) Change detection using Ikonos imagery. In: *Proceedings of the ASPRS 2003 Annual Conference, Anchorage, Alaska, May 5–9*, American Society of Photogrammetry and Remote Sensing, 9 pp.

Hay, G. and G. Castilla (2006) Object-based image analysis: Strengths, weaknesses, opportunities and threats (SWOT). In: *International Conference on Object-based Image Analysis, Salzburg University, Austria, July 4–5*, ISPRS Archives – Volume XXXVI-4/C42, 2006, 8–10.

Hay, G. and G. Castilla (2008) Geographic Object-Based Image Analysis (GEOBIA): A new name for a new discipline. In: *Object-Based Image Analysis: Spatial Concepts for Knowledge-Driven Remote Sensing Applications*, eds Blaschke, T., Lang, S. and Hay, G.J., Berlin: Springer, 75–89.

Hayes, D. J. and S. A. Sader (2001) Comparison of change-detection techniques for monitoring tropical forest clearing and vegetation regrowth in a time series. *Photogrammetric Engineering and Remote Sensing* **67**(9), 1067–1075.

Hecheltjen, A., F. Thonfeld and G. Menz (2014) Recent advances in remote sensing change detection: A review. In: *Land Use and Land Cover Mapping in Europe*, eds Manakos, I. and Braun, M., Berlin: Springer, 145–178.

HEXAGON (2014) ERDAS IMAGINE, http://www.hexagongeospatial.com/products/remote-sensing/erdas-imagine/overview. (last accessed: 26 November 2014).

Hirschmüller, H. (2005) Accurate and efficient stereo processing by semi-global matching and mutual information. In: 2005 *IEEE Computer Society Conference on Computer Vision and Pattern Recognition (CVPR'05)*, 807–814, vol. 2.

Hirschmüller, H. (2008) Stereo processing by semiglobal matching and mutual information. *IEEE Transactions on Pattern Analysis and Machine Intelligence* **30**(2), 328–341.

Howarth, P. J. and G. M. Wickware (1981) Procedures for change detection using Landsat digital data. *International Journal of Remote Sensing* **2**(3), 277–291.

Hu, Y., L. Liu, L. Liu and Q. Jiao (2011) Comparison of absolute and relative radiometric normalization use Landsat time series images. In: *Proc. SPIE*, 800616.

Huang, X. (2013) Building reconstruction from airborne laser scanning data. *Geo-spatial Information Science* **16**(1), 35–44.

Huang, X., L. Zhang and W. Gong (2011) Information fusion of aerial images and LIDAR data in urban areas: Vector-stacking, re-classification and post-processing approaches. *International Journal of Remote Sensing* **32**(1), 69–84.

Huang, X., L. Zhang and T. Zhu (2014) Building change detection from multitemporal high-resolution remotely sensed images based on a morphological building index. *IEEE Journal of Selected Topics in Applied Earth Observations and Remote Sensing* **7**(1), 105–115.

Hussain, M., D. Chen, A. Cheng, H. Wei and D. Stanley (2013) Change detection from remotely sensed images: From pixel-based to object-based approaches. *ISPRS Journal of Photogrammetry and Remote Sensing* **80**(2013), 91–106.

İlsever, M. and C. Unsalan (2012) *Two-dimensional Change Detection Methods: Remote Sensing Applications*. Springer Briefs in Computer Science. Berlin: Springer Science & Business Media, x + 72 pp.

Ingram, K., E. Knapp and J. Robinson (1981) Change detection technique development for improved urbanized area delineation. In: *Technical Memorandum CSC/TM-81*, Computer Sciences Corporation, Silver Springs, Maryland, USA,136 pp.

James, L. A., M. E. Hodgson, S. Ghoshal and M. M. Latiolais (2012) Geomorphic change detection using historic maps and DEM differencing: The temporal dimension of geospatial analysis. *Geomorphology* **137**(1), 181–198.

Jensen, J. R. and D. L. Toll (1982 Detecting residential land-use development at the urban fringe. *Photogrammetric Engineering and Remote Sensing* **48**(4), 629–643.

Jianya, G., S. Haigang, M. Guorui and Z. Qiming (2008) A review of multi-temporal remote sensing data change detection algorithms. *International Archives of the Photogrammetry, Remote Sensing and Spatial Information Sciences* **37**(B7), 757–762.

Johnson, R. D. and E. Kasischke (1998) Change vector analysis: A technique for the multispectral monitoring of land cover and condition. *International Journal of Remote Sensing* **19**(3), 411–426.

Kang, Z. and Z. Lu (2011) The change detection of building models using epochs of terrestrial point clouds. In: *IEEE International Workshop on Multi-Platform/Multi-Sensor Remote Sensing and Mapping Xiamen, China, 10–12 January*, 1–6.

Karras, G. E. and E. Petsa (1993) DEM matching and detection of deformation in close-range photogrammetry without control. *Photogrammetric Engineering and Remote Sensing* **59**(9), 1419–1424.

Knudsen, T. and B. P. Olsen (2003) Automated change detection for updates of digital map databases. *Photogrammetric Engineering and Remote Sensing* **69**(11), 1289–1296.

Kohonen, T. (1982) Self-organized formation of topologically correct feature maps. *Biological Cybernetics* **43**(1), 59–69.

Krauss, T., P. d'Angelo, M. Schneider and V. Gstaiger (2013) The fully automatic optical processing system CATENA at DLR. *International Archives of the Photogrammetry, Remote Sensing and Spatial Information Sciences* vol. XL-1/W1, 177–181.

Lambin, E. F. and D. Ehrlich (1997) Land-cover changes in sub-Saharan Africa (1982–1991): Application of a change index based on remotely sensed surface temperature and vegetation indices at a continental scale. *Remote Sensing of Environment* **61**(2), 181–200.

LeCun, Y., Y. Bengio and G. Hinton (2015) Deep learning. *Nature* **521**(7553), 436–444.

Lillesand, T. M., R. W. Kiefer and J. W. Chipman (2004) *Remote Sensing and Image Interpretation. 5th Edition*. Chichester: John Wiley & Sons Ltd, 804 pp.

Listner, C. and I. Niemeyer (2011) Recent advances in object-based change detection. In: *2011 IEEE International Geoscience and Remote Sensing Symposium (IGARSS),* pp. 110–113.

Liu, Z., J. Zhang, Z. Zhang and H. Fan (2003) Change detection based on DSM and image features in urban areas. *Geo-spatial Information Science* **6**(2), 35–41.

Longbotham, N., F. Pacifici, T. Glenn, A. Zare, M. Volpi, D. Tuia, E. Christophe, J. Michel, J. Inglada and J. Chanussot (2012) Multi-modal change detection, application to the detection of flooded areas: Outcome of the 2009–2010 data fusion contest. *IEEE Journal of Selected Topics in Applied Earth Observations and Remote Sensing* **5**(1), 331–342.

Longley, P. A. (2002) Geographical information systems: Will developments in urban remote sensing and GIS lead to 'better'urban geography? *Progress in Human Geography* **26**(2), 231–239.

Lu, D., P. Mausel, E. Brondızio and E. Moran (2002) Change detection of successional and mature forests based on forest stand characteristics using multitemporal TM data in Altamira, Brazil. In: *XXII FIG International Federation of Surveyors, ACSM–ASPRS Annual Conference Proceedings, Washington, DC, USA*, 19–26.

Lu, D., P. Mausel, E. Brondizio and E. Moran (2004) Change detection techniques. *International Journal of Remote Sensing* **25**(12), 2365–2401.

Maas, H.-G. (2000) Least-squares matching with airborne laserscanning data in a TIN structure. *International Archives of Photogrammetry and Remote Sensing* **33**(B3/1; PART 3), 548–555.

Malila, W. A. (1980) Change vector analysis: An approach for detecting forest changes with Landsat. In: *LARS(Laboratory for Applications of Remote Sensing) Symposia*, Paper 385, 326–335.

Martha, T. R., N. Kerle, V. Jetten, C. J. van Westen and K. V. Kumar (2010) Landslide volumetric analysis using Cartosat-1-derived DEMs. *IEEE Geoscience and Remote Sensing Letters* **7**(3), 582–586.

Mas, J.-F. (1999) Monitoring land-cover changes: A comparison of change detection techniques. *International Journal of Remote Sensing* **20**(1), 139–152.

Matikainen, L., J. Hyyppä, E. Ahokas, L. Markelin and H. Kaartinen (2010) Automatic detection of buildings and changes in buildings for updating of maps. *Remote Sensing* **2**(5), 1217–1248.

Mayer, H. (1999) Automatic object extraction from aerial imagery: A survey focusing on buildings. *Computer Vision and Image Understanding* **74**(2), 138–149.

Menderes, A., A. Erener and G. Sarp (2015) Automatic detection of damaged buildings after earthquake hazard by using remote sensing and information technologies. *Procedia Earth and Planetary Science* **15**, 257–262.

Metternicht, G. (1999) Change detection assessment using fuzzy sets and remotely sensed data: An application of topographic map revision. *ISPRS Journal of Photogrammetry and Remote Sensing* **54**(4), 221–233.

Mitchell, H. and R. Chadwick (1999) Digital photogrammetric concepts applied to surface deformation studies. *Geomatica* **53**(4), 405–414.

Moosavi, S. V. and R. Qin (2012) A new automated hierarchical clustering algorithm based on emergent self-organizing maps. In: *16th International Conference on Information Visualisation,* Montpellier, France, 2012, pp. 264–269, doi: 10.1109/IV.2012.52.

Morgan, G. L. K., J. G. Liu and H. Yan (2010) Precise subpixel disparity measurement from very narrow baseline stereo. *IEEE Transactions on Geoscience and Remote Sensing* **48**(9), 3424–3433.

Muchoney, D. M. and B. N. Haack (1994) Change detection for monitoring forest defoliation. *Photogrammetric Engineering and Remote Sensing* **60**(10), 1243–1251.

Munyati, C. (2004) Use of principal component analysis (PCA) of remote sensing images in wetland change detection on the Kafue Flats, Zambia. *Geocarto International* **19**(3), 11–22.

Murakami, H., K. Nakagawa, H. Hasegawa, T. Shibata and E. Iwanami (1999) Change detection of buildings using an airborne laser scanner. *ISPRS Journal of Photogrammetry and Remote Sensing* **54**(2), 148–152.

Nebiker, S., N. Lack and M. Deuber (2014) Building change detection from historical aerial photographs using dense image matching and object-based image analysis. *Remote Sensing* **6**(9), 8310–8336.

Nelson, R. (1983) Detecting forest canopy change due to insect activity using Landsat MSS. *Photogrammetric Engineering and Remote Sensing* **49**, 1303–1314.

Nemmour, H. and Y. Chibani (2006) Multiple support vector machines for land cover change detection: An application for mapping urban extensions. *ISPRS Journal of Photogrammetry and Remote Sensing* **61**(2), 125–133.

Nex, F. and F. Remondino (2014) UAV for 3D mapping applications: a review. *Applied Geomatics* **6**(1), 1–15.

Niemeyer, I., P. R. Marpu and S. Nussbaum (2008) Change detection using object features. *Object-Based Image Analysis: Spatial Concepts for Knowledge-Driven*, eds Blaschke, T., Lang, S. and Hay, G.J. Lecture Notes in Geoinformation and Cartography. Berlin: Springer, 185–201.

Olsen, B. P. (2004) Automatic change detection for validation of digital map databases. *International Archives of Photogrammetry and Remote Sensing* **34**(Part B2), 569–574.

Olsen, B. P. and T. Knudsen (2005) Automated change detection for validation and update of geodata. In: 6th Geomatic Week Conference, Barcelona, Spain.

Pacifici, F., F. Del Frate, C. Solimini and W. J. Emery (2007) An innovative neural-net method to detect temporal changes in high-resolution optical satellite imagery. *IEEE Transactions on Geoscience and Remote Sensing* **45**(9), 2940–2952.

Pacifici, F., M. Chini and W. J. Emery (2009) A neural network approach using multi-scale textural metrics from very high-resolution panchromatic imagery for urban land-use classification. *Remote Sensing of Environment* **113**(6), 1276–1292.

Pang, S., X. Hu, Z. Wang and Y. Lu (2014) Object-based analysis of airborne LiDAR data for building change detection. *Remote Sensing* **6** (11), 10733–10749.

Paolini, L., F. Grings, J. A. Sobrino, J. C. Jiménez Muñoz and H. Karszenbaum (2006) Radiometric correction effects in Landsat multi-date/multi-sensor change detection studies. *International Journal of Remote Sensing* **27**(4), 685–704.

Parra, G. A., M.-C. Mouchot and C. Roux (1996) A multitemporal land-cover change analysis tool using change vector and principal components analysis. In: *1996 International IEEE Geoscience and Remote Sensing Symposium, 'Remote Sensing for a Sustainable Future',* pp. 1753–1755, vol. 1.

Pilgrim, L. (1996) Robust estimation applied to surface matching. *ISPRS Journal of Photogrammetry and Remote Sensing* **51**(5), 243–257.

Pix4D. (2017) Pix4D, http://pix4d.com/ (last accessed: June 09 2017).

Pollard, T. and J. L. Mundy (2007) Change detection in a 3-d world. In: *2007 IEEE Computer Society Conference on Computer Vision and Pattern Recognition*, pp. 1–6.

Qin, R. (2014a) Change detection on LOD 2 building models with very high resolution spaceborne stereo imagery. *ISPRS Journal of Photogrammetry and Remote Sensing* **96**(2014), 179–192.

Qin, R. (2014b) An object-based hierarchical method for change detection using unmanned aerial vehicle images. *Remote Sensing* **6**(9), 7911–7932.

Qin, R.( 2016) RPC Stereo Processor (RSP) – A software package for digital surface model and orthophoto generation from satellite stereo imagery. *ISPRS Annals of Photogrammetry, Remote Sensing and Spatial Information Sciences* **III**(1), 77–82. DOI: 10.5194/isprs-annals-III-1-77-2016.

Qin, R. and W. Fang (2014) A hierarchical building detection method for very high resolution remotely sensed images combined with DSM using graph cut optimization. *Photogrammetry Engineering and Remote Sensing* **80**(8), 37–48.doi: 10.14358/PERS.80.9.000

Qin, R. and A. Gruen (2014) 3D change detection at street level using mobile laser scanning point clouds and terrestrial images. *ISPRS Journal of Photogrammetry and Remote Sensing* **90**(2014), 23–35.

Qin, R., J. Gong, H. Li and X. Huang (2013) A coarse elevation map-based registration method for super-resolution of three-line scanner images. *Photogrammetric Engineering and Remote Sensing* **79**(8), 717–730.

Qin, R., X. Huang, A. Gruen and G. Schmitt (2015) Object-based 3-D building change detection on multitemporal stereo images. *IEEE Journal of Selected Topics in Applied Earth Observations and Remote Sensing* **5**(8), 2125–2137. 10.1109/JSTARS.2015.2424275

Qin, R., J. Tian and P. Reinartz (2016a) 3D change detection – Approaches and applications. *ISPRS Journal of Photogrammetry and Remote Sensing* **122,** 41–56.

Qin, R., J. Tian and P. Reinartz (2016b) Spatiotemporal inferences for use in building detection using series of very-high-resolution space-borne stereo images. *International Journal of Remote Sensing* **37**(15), 3455–3476. DOI: 10.1080/01431161.2015.1066527.

Radke, R. J., S. Andra, O. Al-Kofahi and B. Roysam (2005) Image change detection algorithms: A systematic survey. *IEEE Transactions on Image Processing* **14**(3), 294–307.

Ram, B. and A. Kolarkar (1993) Remote sensing application in monitoring land-use changes in arid Rajasthan. *International Journal of Remote Sensing* **14**(17), 3191–3200.

Remondino, F. and S. El-Hakim (2006) Image-based 3D modelling: A review. *The Photogrammetric Record* **21**(115), 269–291.

Remondino, F., M. G. Spera, E. Nocerino, F. Menna and F. Nex (2014) State of the art in high density image matching. *The Photogrammetric Record* **29**(146), 144–166.

Rogan, J., J. Franklin and D. A. Roberts (2002) A comparison of methods for monitoring multitemporal vegetation change using Thematic Mapper imagery. *Remote Sensing of Environment* **80**(1), 143–156.

Rosenholm, D. and K. Torlegard (1988) Three-dimensional absolute orientation of stereo models using digital elevation models. *Photogrammetric Engineering and Remote Sensing* **54,** 1385–1389.

Rottensteiner, F., J. Trinder, S. Clode and K. Kubik (2007) Building detection by fusion of airborne laser scanner data and multi-spectral images: Performance evaluation and sensitivity analysis. *ISPRS Journal of Photogrammetry and Remote Sensing* **62**(2), 135–149.

Roy, D. P., J. Ju, P. Lewis, C. Schaaf, F. Gao, M. Hansen and E. Lindquist (2008) Multi-temporal MODIS–Landsat data fusion for relative radiometric normalization, gap filling, and prediction of Landsat data. *Remote Sensing of Environment* **112**(6), 3112–3130.

Sasagawa, A., E. Baltsavias, S. K. Aksakal and J. D. Wegner (2013) Investigation on automatic change detection using pixel-changes and DSM-changes with ALOS-PRISM triplet images. *ISPRS-International Archives of the Photogrammetry, Remote Sensing and Spatial Information Sciences* **1**(2), 213–217.

Sasagawa, A., K. Watanabe, S. Nakajima, K. Koido, H. Ohno and H. Fujimura (2008) Automatic change detection based on pixel-change and DSM-change. *The International Archives of the Photogrammetry, Remote Sensing and Spatial Information Sciences* **37**(Part B7), 1645–1650.

Schindler, G. and F. Dellaert (2010) Probabilistic temporal inference on reconstructed 3d scenes. In: *2010 IEEE Conference on Computer Vision and Pattern Recognition*, pp. 1410–1417.

Seitz, S. M., B. Curless, J. Diebel, D. Scharstein and R. Szeliski (2006) A comparison and evaluation of multi-view stereo reconstruction algorithms. In: *2006 IEEE Conference on Computer Society Conference* on *Computer Vision and Pattern Recognition, (CVPR'06),* pp. 519–528.

Siebert, S. and J. Teizer (2014) Mobile 3D mapping for surveying earthwork projects using an Unmanned Aerial Vehicle (UAV) system. *Automation in Construction* **41,** 1–14.

Singh, A. (1986) Change detection in the tropical forest environment of northeastern India using Landsat. In: *Remote Sensing and Tropical Land Management*, eds Eden, M. J. and Parry, J.T. Chichetser: Wiley, 237–254.

Singh, A. (1989) Digital change detection techniques using remotely-sensed data. *International Journal of Remote Sensing* **10**(6), 989–1003.

Sohn, G. and I. Dowman (2007) Data fusion of high-resolution satellite imagery and LiDAR data for automatic building extraction. *ISPRS Journal of Photogrammetry and Remote Sensing* **62**(1), 43–63.

Stepper, C., C. Straub and H. Pretzsch (2015) Assessing height changes in a highly structured forest using regularly acquired aerial image data. *Forestry* **88**(3), 304–316.

Strong, A. E. (1974) Remote sensing of algal blooms by aircraft and satellite in Lake Erie and Utah Lake. *Remote Sensing of Environment* **3**(2), 99–107.

Taneja, A., L. Ballan and M. Pollefeys (2011) Image based detection of geometric changes in urban environments. In: *2011 IEEE International Conference on Computer Vision*, pp. 2336–2343.

Tao, H., H. S. Sawhney and R. Kumar (2001) A global matching framework for stereo computation. In: *Proceedings Eighth IEEE International Conference on Computer Vision. ICCV 2001.* pp. 532–539.

Teo, T.-A. and T.-Y. Shih (2013) Lidar-based change detection and change-type determination in urban areas. *International Journal of Remote Sensing* **34**(3), 968–981.

Terrasolid. (2013) Terrasolid – Software for airborne and mobile LiDAR and image processing, http://www.terrasolid.fi/. (Accessed 11 September, 2013)

Tewkesbury, A. P., A. J. Comber, N. J. Tate, A. Lamb and P. F. Fisher (2015) A critical synthesis of remotely sensed optical image change detection techniques. *Remote Sensing of Environment* **160,** 1–14.

Tian, J., H. Chaabouni-Chouayakh, P. Reinartz, T. Krauß and P. d'Angelo (2010) Automatic 3D change detection based on optical satellite stereo imagery. *International Archives of Photogrammetry, Remote Sensing and Spatial Information Sciences* **38**(Part 7B), 586–591.

Tian, J., A. A. Nielsen and P. Reinartz (2014) Improving change detection in forest areas based on stereo panchromatic imagery using kernel MNF. *IEEE Transactions on Geoscience and Remote Sensing* **52**(11), 7130–7139.

Tian, J., P. Reinartz, P. d'Angelo and M. Ehlers (2013) Region-based automatic building and forest change detection on Cartosat-1 stereo imagery. *ISPRS Journal of Photogrammetry and Remote Sensing* **79,** 226–239.

Tornabene, T., G. Holzer, S. Lien and N. Burris (1983) Lipid composition of the nitrogen starved green alga *Neochloris oleoabundans. Enzyme and Microbial Technology* **5**(6), 435–440.

Triggs, B., P. F. McLauchlan, R. I. Hartley and A. W. Fitzgibbon (2000) *Bundle Adjustment: A Modern Synthesis*. Berlin: Springer, 298–372.

Trimble (2014) eCognition, http://www.ecognition.com/. (last accessed: 26 November 2014)

Turker, M. and B. Cetinkaya (2005) Automatic detection of earthquake-damaged buildings using DEMs created from pre-and post-earthquake stereo aerial photographs. *International Journal of Remote Sensing* **26**(4), 823–832.

Walter, V. (2004) Object-based classification of remote sensing data for change detection. *ISPRS Journal of Photogrammetry and Remote Sensing* **58**(3), 225–238.

Wang, L. (2005) *Support Vector Machines: Theory and Applications*. Berlin: Springer, pp. 431.

Waser, L., E. Baltsavias, K. Ecker, H. Eisenbeiss, E. Feldmeyer-Christe, C. Ginzler, M. Küchler and L. Zhang (2008) Assessing changes of forest area and shrub encroachment in a mire ecosystem using digital surface models and CIR aerial images. *Remote Sensing of Environment* **112**(5), 1956–1968.

Waser, L. T., E. Baltsavias, H. Eisenbeiss, C. Ginzler, A. Gruen, M. Kuechler and P. Thee (2007) Change detection in mire ecosystems: Assessing changes of forest area using airborne remote sensing data. In: *Remote Sensing and Spatial Information Sciences.* International Archives of Photogrammetry, vol. 36, pp. 313–318

Wu, B., Y. Zhang and Q. Zhu (2012) Integrated point and edge matching on poor textural images constrained by self-adaptive triangulations. *ISPRS Journal of Photogrammetry and Remote Sensing* **68,** 40–55.

Yang, Q. (2012) A non-local cost aggregation method for stereo matching. In: *2012 IEEE Conference on Computer Vision and Pattern Recognition (CVPR)*, pp. 1402–1409.

Yang, X. and C. Lo (2000) Relative radiometric normalization performance for change detection from multi-date satellite images. *Photogrammetric Engineering and Remote Sensing* **66**(8), 967–980.

Ye, S., D. Chen and J. Yu (2016) A targeted change-detection procedure by combining change vector analysis and post-classification approach. *ISPRS Journal of Photogrammetry and Remote Sensing* **114,** 115–124.

Young, A. P., M. Olsen, N. Driscoll, R. Flick, R. Gutierrez, R. Guza, E. Johnstone and F. Kuester (2010) Comparison of airborne and terrestrial lidar estimates of seacliff erosion in southern California. *Photogrammetric Engineering & Remote Sensing* **76**(4), 421–427.

Yu, Q., P. Gong, N. Clinton, G. Biging, M. Kelly and D. Schirokauer (2006) Object-based detailed vegetation classification with airborne high spatial resolution remote sensing imagery. *Photogrammetric Engineering & Remote Sensing* **72**(7), 799–811.

Zang, D. and G. Zhou (2007) Area spatial object co-registration between imagery and GIS data for spatial-temporal change analysis. In: *2007 IEEE International Geoscience and Remote Sensing Symposium*. pp. 2597–2600.

Zhang, Q., R. Qin, X. Huang, Y. Fang and L. Liu (2015) Classification of ultra-high resolution orthophotos combined with DSM using a dual morphological top hat profile. *Remote Sensing* **7**(12), 16422–16440.

Zhu, L., H. Shimamura, K. Tachibana, Y. Li and P. Gong (2008) Building change detection based on object extraction in dense urban areas. *International Archives of Photogrammetry, Remote Sensing and Spatial* Information Sciences **37**(Part B7), 905–908.

# Chapter 9
# The Topographic Landscape Model of Switzerland swissTLM$^{3D}$

*Patrick Aeby, Tobias Kellenberger, Liam O'Sullivan, Emanuel Schmassmann, André Streilein and Michael Zwick*

Digital 3D geo-information has been available in various forms for many years. Within the national mapping community 3D geo-information was first used to create contour lines on topographic maps and in the private sector to produce small-extent digital terrain models (DTM) to support engineering projects. Acquiring raw 3D data and using it to construct 3D geo-information has been expensive in the past. Consequently, 3D geo-information has usually only been produced for limited areas on a project-by-project basis (DTMs and point clouds have been produced at national coverages).

In recent years, the introduction of more efficient LiDAR and imaging cameras and related processing solutions has significantly reduced the cost of acquiring and extracting 3D geo-information. Platforms including earth observation satellites, airplanes, UAVs (drones) and vehicles are competing in the high-resolution data acquisition market. Pattern recognition software to process images for the creation of point clouds, vector polygons and land use classifications has similarly become more capable.

Consequently, the modelling of our world in 3D is finding an ever-increasing array of applications. National mapping agencies (NMAs) are looking to respond to this opportunity by re-orienting their data operations and processes to create 3D geo-information by default and, as a result, to re-position their products and services to monitor rapid and radical changes in the expectations of the public and private sector user base.

## 9.1 The Topographic Landscape Model of Switzerland, TLM

TLM is the basic landscape model for Switzerland (Figure 9.1) and covers the whole country. It is a seamless Geographic Information System (GIS) containing 3D primary geometry without generalisation and with an accuracy of better than 1 meter in $x$, $y$ and $z$ coordinates for most objects, and even better accuracy for roofs – of the order of 0.5 meter or better. The TLM consists of nine topics that contain several dependent datasets previously maintained separately, such as boundaries, names and terrain (Figure 9.2).

It also has a revised extended data model to meet the needs of reference partners like road and water authorities. The approach realised with TOPGIS makes updates of

Figure 9.1 Visualisation of TLM data in the region of Interlaken.

TLM data a lot easier and more frequent than the present 6-year map production cycle (certainly, often-changing objects such as roads and buildings will, for example, be updated annually in a so-called top up-to-dateness). Reference partners (the cadastre, road and rail authorities among others) deliver their up-to-date data for rapid insertion into the TLM in order to maintain top up-to-dateness. All objects are stored in 3D and

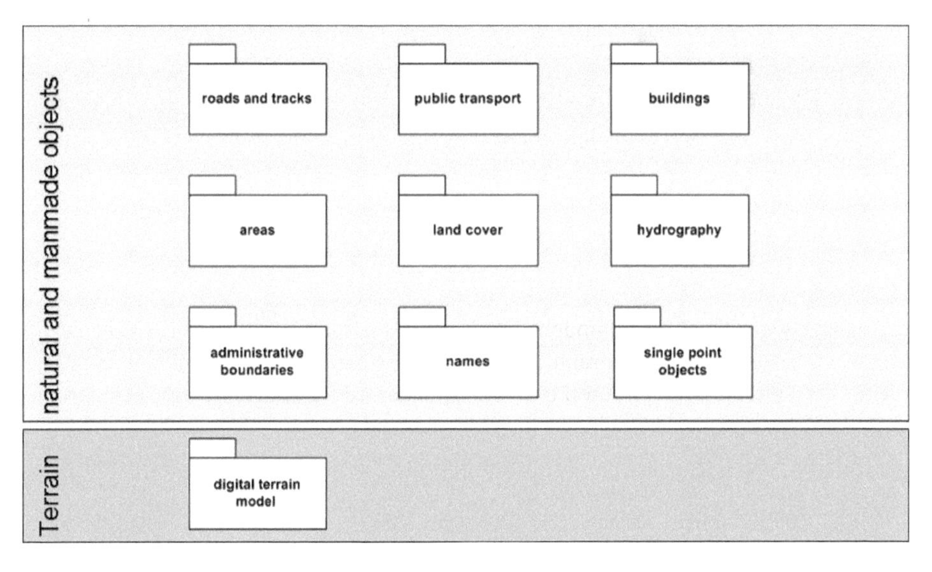


Figure 9.2 TLM has 9 topic classes and bases on a DTM.

have *x*, *y* and *z* coordinates, which also means that objects in the TLM should maintain consistency with the DTM. The TLM is not a product: it is, rather, the basis for a wide variety of products including the derived cartographic basemaps.

The DTM covers the entire surface of Switzerland to a high degree of accuracy. It is derived principally from LiDAR data and includes mass point data, breaklines and polygons as well as attributes. The DTM is stored in the ESRI Geodatabase Terrain format and is updated parallel to the updating of the TLM.

TLM and DTM data are based on the Swiss local reference system (CH1903+ (LV95)) and the height system LHN95. Transformations from other reference systems and height systems are applied on-the-fly. The user doesn't need to worry about coordinate systems.

### 9.1.1 A short look in the past

The ever-increasing demand for better, more up-to-date and diverse digital data requires the rapid updating of these geo-products. Until 2008 several of these major products – such as the cartographic basemap 1:25,000 (LK25), the vector landscape model VECTOR25 as well as the SWISSNAMES Database (SND) – were updated independently of each other. The updating of these products has been partially analogue, or raster- and vector-based using a range of heterogeneous updating systems, formats, methods and tools. Indeed, some products such as the map contour digitized Digital Height Model (DHM25) and even a LiDAR DTM product (DTM-AV) had not been updated regularly – or at all. The production lines were complex, involving lots of individual solutions. The uniqueness of each system, program and individual production line also meant that staff could not easily be moved from one part of the production to another without considerable retraining.

Motivation for the TLM was to establish a central 3D dataset from which all vector products could be produced. The updating should take place over a shorter length of time (flight to product delivery time) with a top up-to-dateness and yearly updating for certain important levels such as roads and buildings.

### 9.1.2 Product updating before the introduction of TLM

In the past, prior to the introduction of the TOPGIS system for the acquisition of TLM data, raster LK25 maps were updated on a sheet-by-sheet basis every 6 years from analogue 1:30,000 scale photography using analytical photogrammetry. Mutations reported by the public were accumulated and printed on a special revision map. The observed changes were captured stereoscopically into an object-coded vector file superimposed on the raster map. Stereo vector superimposition was not available. After verification in the field, the vector updates were delivered to the cartography department in 2D CAD format for integration into the 1:25,000 pixel maps. The third dimension gathered during the photogrammetric stage was not used in the production process; it was a purely 2D updating.

VECTOR25 was a seamless 2D digital landscape model of Switzerland comprising of about 5 million objects showing their position, form and attributes in a GIS. It was created by digitising the LK25 maps and had the same accuracy (3–8 m) and revision year as the maps, but it was published later and was pretty out of date by the time it came out. For historical reasons it was derived from the LK25 maps and thus was limited to a map-based content, but unfortunately it did not contain the entire LK25 content. Developments over the years led to this dual updating process, which was rather cumbersome for the users because of heterogeneous production environments, data exchange via interfaces, etc. It was considered wasteful of resources and also created an unnecessary delay in that it was impossible to update VECTOR25 more frequently than its map source i.e. once every 6 years. As a result, customers received data that were already old and which could not be geometrically improved before the maps were updated.

The database of Swiss-names (SWISSNAMES) had been used for updating the place names in all the Swiss map scales and had therefore been map based. Importantly, the names only had a single geographical reference point, whereas, in the TLM, the names would be assigned to an area. Field updating has come a long way since the analogue maps used to be taken into the field for the verification.

### 9.1.3 What needed to be done?

The various product-updating methods were developed independently over the years and slowly started to outlive their usefulness. The dependency of VECTOR25 and SWISSNAMES on the map-updating process needed to be uncoupled. It was necessary to eliminate this independent, uncoordinated and parallel updating and replace it by a single updating operation from a central dataset from which all products could later be derived. Realising this goal meant that a large part of the production flow needed to be reversed. The production of the data would be shortened as the updating could take place independently and more frequently than the 6-year map production cycle.

### 9.1.4 TOPGIS – the production infrastructure

TOPGIS is a modern database system for the capture, editing, management and storage of the TLM. The associated LiDAR-based Digital Terrain Model (DTM) is also edited using TOPGIS. TOPGIS is completely based on ArcGIS including desktop, server, and mobile GIS components. It comprises specially in-built extensions for data capture and management. It has fully integrated photogrammetric features through the extended Stereo Analyst Extension for ArcGIS (SAFA) from Hexagon. The main source for the TLM data is aerial images from the Federal Office of Topography (swisstopo). The imagery is flown in a 3-year cycle over the whole country. From 2008 to 2017 the image resolution was 25 cm/50 cm, then it increased to 10 cm/25 cm, depending on the altitude. Although the TLM is 3D, not all objects are edited in stereo: indeed, many objects are actually 2.5D in that the Z value is the same as that of the underlying DTM. TOPGIS supports the capture, editing and management of GIS data (TLM and DTM) in several modes including 3D and 2.5D from the following clients:

- 3D client (stereo) for objects which 'form or shape' the terrain, such as waterways, roads, railways, and wherever object heights are captured, such as roofs, electric poles, bridges, etc.

- 2.5D client (mono-plotting) for objects that 'sit' on the terrain, such as hiking trails, land cover, boundaries, etc.
- DTM client (stereo) for DTM editing.
- Mobile client for 3D field editing.

These clients may work together or alone: for example, an operator may use the 3D client and the 2.5D client on the same PC, as they interact seamlessly. An object such as a road can be started in 3D and then continued in 2.5D from a different image source (i.e. a shaded terrain relief), where for example the ground is no longer visible in 3D (i.e. in a dense forest), and finally completed in 3D mode. The operator simply switches from the 3D screen to the 2D screen; the object is always visible on both screens. 3D clients are always paired with a 2.5D client, but a 2.5D client may exist independently for non-stereo capture such as attribute collection or name collection.

## 9.2 Migration VECTOR25 to TLM

The TLM DB was not created from scratch but was initially populated, mainly but not exclusively, by transforming the existing 2D VECTOR25 dataset. This 2D dataset had a map-based schema and needed to undergo the following transformations before it could be put into the TLM DB:

- Schema transformation from the VECTOR25 schema to the TLM 'real-world' oriented schema including the generation of new object classes and populating it with new attributes.
- Coordinate transformation from the old LV03 and LN02 to the new LV95 and LHN95 systems.
- Z awareness by transforming from 2D to 2.5D by the addition of the third dimension to the data from DTM sources.
- Transformation of point, line and polygon geometries to newly defined geometries including the creation of standard-sized polygons and symbols from points.

### 9.2.1 Geometric de-generalisation of the data

Once the data were successfully transformed and inserted into the DB, the laborious task of de-generalisation and re-engineering of the TLM data could begin. This meant that the former map-based geometry of, for example, the road network (see Figure 9.3) had to be improved using aerial images (see Figures 9.4 and 9.5). All data were de-generalised or partially restituted; buildings were all newly captured with stereo-restitution in 3D. The operations were completed at the beginning of 2018. The buildings were initially kept as polygons, maintaining their valuable names, usage attributes and object classes information. In order to maintain stable universally unique identifiers (UUIDs), some objects, like roads, were geometrically improved, but if this proved too difficult or impossible to perform, they were re-captured completely. However, not all the data could be improved: in particular,

Figure 9.3 1:25,000 map with VECTOR25 road axes digitised directly from the pixel map.

Figure 9.4 Orthophoto with map-based VECTOR25 road axes superimposed.

those roads, tracks and buildings that remained hidden in densely forested areas. Such data that had previously been painstakingly captured using field surveys and cannot be verified in the aerial imagery due to occlusions, are still (to this day) assumed as correct, as new field verification would be too expensive.

Figure 9.5 Orthophoto with map-based VECTOR25 and the new 3D TLM road axes superimposed.

### 9.2.2 Densifying the data (add all missing objects and attributes)

As the previous VECTOR25 data were map-based a lot of objects existing in the real world were not in the dataset, because they were originally covered only by names or cartographically generalised. Hence, thousands of paths and shorter road segments, hundreds of kilometers of forest edges and smaller 'unimportant' map objects that were never included in the VECTOR25 schema had to be added manually, thus improving the existing data temporally while guaranteeing the correctness of the topology.

### 9.2.3 Add 3D to the data

A first approach for adding the third dimension to the data, at least for all objects 'laying on the ground', was to project them on to the existing DTM. Afterwards, the Z-values were, wherever required, corrected manually with stereoscopic editing using aerial images. All the objects 'not laying on the ground', such as buildings, bridges, lake edges, monuments, walls, dams, etc., had to be captured in 3D. For example, the roofs of about 3.7 million buildings of the entire country were captured and the buildings were modelled as volumes.

Effectively, all the improvements, i.e. geometric de-generalisation, data densification and adding the correct 3D, were carried out simultaneously with the data updating. This process was not performed for all objects at the same time. Firstly, the public transportation network (railway lines, tram lines, cable cars, etc.) and the roads were improved, because they were the most urgently needed. Afterwards, there was a special program for the buildings, then for Land cover, hydrography and land use. Currently (April 2023) swissTLM3D is available for all of Switzerland and the Principality of Liechtenstein.

### 9.2.4 Lessons learned

Lessons learned are:

- Do not try to change your whole production scheme at once. Change it gradually: for example, start with geometric improvements ASAP as this is the most time consuming!
- Test your new production and data model with an extensive pilot project (old and new production in parallel).
- DTM should ideally be correct before 2.5D capture begins.
- If possible, finish the migration before production starts (to have the basis for increments, consistency, etc.).
- Correct network data first (roads, rail, hydrology) and remove ongoing conflicts with other data.
- Automate production wherever possible or make provisions for increasing the personnel capable of processing 3D stereo!

## 9.3 The $A^4C^4$ quality requirements for swissTLM$^{3D}$

Advances in technology, the digitisation of society, changes in communication and increased mobility have a growing influence on the value and the quality requirements of geographic information (Sulaiman and Gudmundsdottir, 2013). The mapping agencies face an augmented demand for more detailed and complex data, shorter update cycles and accelerated speed to market. Topographic data acquisition is gradually leaving the controlled in-house environment of mapping agencies and moving towards the crowds of external experts and the public (Fonte *et al.*, 2015). This trend raises directly the awareness of data quality. Many publications discuss quality requirements and principles for topographic data, or more generally for GIS data (Chapman, 2005; Shi *et al.*, 2002; Zhang and Goodchild, 2002). Guidelines for implementing the ISO 19100 standard series assist the mapping agencies on the technical aspects of data quality assurance such as data model, consistency, precision, metadata, etc. (Jakobsson and Giversen, 2007). However, they do not cover the whole range of quality aspects that swisstopo has to address as an authoritative public body in relation to its data. Furthermore, most of the known quality principles and requirements address the 2D data aspects and not the 3D, such as required for the swissTLM$^{3D}$ landscape model.

In the swisstopo portfolio the quality assurance for all topographic data, independent of the origin or dimensionality, relies on eight elements, named '$A^4C^4$ quality requirements' (Figure 9.6).

| The swisstopo **A$^4$C$^4$** quality requirements for topographic data | |
|---|---|
| **A**uthority | (data source, production) |
| **A**ccuracy | (spatial location and precision) |
| **A**vailability | (data access) |
| **A**ctuality | (validity period of data) |
| **C**ompleteness | (controlled by surveying rules) |
| **C**overage | (nationwide homogeneity, not hotspots) |
| **C**onsistency | (data contradictoriness and coherence) |
| **C**orrectness | (comparison model and landscape) |

Figure 9.6 The A$^4$C$^4$ quality requirements for topographic data.

*Authority*: Authoritative datasets such as the swissTLM$^{3D}$ landscape model require that the data sources are managed by authoritative entities (mapping agencies, legal bodies, trusted experts, etc.). Then, the validated or trusted data are respected as an authoritative dataset. The data sources can be centralised, distributed or federated. Swisstopo acts in the production environment of the landscape model as an authoritative producer (mapping) as well as a trusted integrator (validating) of external data.

*Accuracy*: Accuracy refers first to the closeness of the measured values to the spatial location of the real mapped objects. For each vector element in the landscape model, we hold a positional error range in all directions (x, y, z). The precision or resolution of the landscape model considers errors due to mapping techniques such as the geometric and numerical resolution of the reference (sampling distance of digital imagery) or individual human capabilities (operator peculiarity).

*Availability*: Users very often identify a source for some target data elements, and then they find out that the dataset is off limits or contains protected information. Access to all layers in a landscape model is an important demand for customers. The dataset must be free of elements from foreign owners and, in the best case, open to the public.

*Actuality*: Actuality is a trade-off between timeliness and data consistency. Customers expect an increasing timeliness for the topographic landscape model and anticipate a granted data consistency. One of the most essential items is indicating the validity period, date and even time of every single landscape model feature.

*Completeness*: Data completeness is the measure of the totality of features and attributes in the landscape model. To achieve completeness in the swissTLM$^{3D}$, swisstopo applies feature-specific surveying rules, requires complete and homogeneous coverage over the entire country, and maps the features based on standardised procedures.

*Coverage*: The unique and complete swissTLM$^{3D}$ dataset covers the entire country with a high degree of quality and homogeneity for all mapping elements. In contrast to volunteered geographical information datasets, such as OpenStreetMap (OSM), the landscape model guarantees the same quality and completeness for all layers over the entire country of Switzerland. The fitness for purpose approach of OSM is in contrast with the legal requirements of authoritative datasets.

*Consistency*: Technical, logical and temporal consistency between all layers is required for each release of swissTLM$^{3D}$. Inherent in the claim of high consistency quality for the landscape model is that all data elements are without contradiction and are coherent. Technical consistency demands on data structures and data acquisition tools together with logical (content) consistency stresses conformal coherence in the dataset (e.g. geometry, attributes, spatial and attributive relations and precision, etc.). Quality checks for consistency are common for 2D topographic data but *still underdeveloped in 3D.*

*Correctness*: The quality requirement of correctness relates first to a proper representation of the (model) specification for each mapped feature and second to an accurate representation of the reality. Quality checks assert the correctness of geometric, attributive and semantic rules of data acquisition and modelling. Most of the landscape models are driven by geometry or discrete vector objects. 3D representation is a good choice for manmade objects because it fits the common understanding of reality. On the other hand, for natural objects with soft or unclear borders, landscape models have to evolve from being geometry-driven to being object-driven or even continua-based models.

## 9.4 Applications

### 9.4.1 Solar

In the topographical landscape model, photogrammetrically generated 3D buildings played a central role for the creation of the Swiss-wide solar potential cadastre. Thanks to the three-dimensional building information from swisstopo, it was possible to realise the interactive applications www.sonnendach.ch and www.sonnenfassade.ch. With the aid of these web pages, the solar energy potential of rooftops and façades in Switzerland can be calculated in an uncomplicated way. This joint project of various federal offices (Federal Office for Energy, Federal Office of Topography and Federal Office of Meteorology and Climatology) contributes to the promotion of a sustainable energy supply in Switzerland and thus also to a successful implementation of the National Energy Strategy 2050.

#### 9.4.1.1 Complex geo-analysis

There are complex and multi-level geo-analysis algorithms behind the easy-to-use mobile applications (Figure 9.7), which would not have been feasible without the high-resolution and high-quality three-dimensional geodata from swisstopo. The building data include the inclination, orientation and extent of the roof surfaces. These are essential input for the precise calculation of the solar energy potential of a roof. Combining this information with the building location in three-dimensional space forms the basis for reliable shade/shadow analysis. Shading should be taken into account for realistic and suitable solar systems assessment on roofs or façades. As a next step, satellite-based radiation data from MeteoSwiss was transformed into a spatial radiation pattern and integrated into the geo-analysis. In order to show how

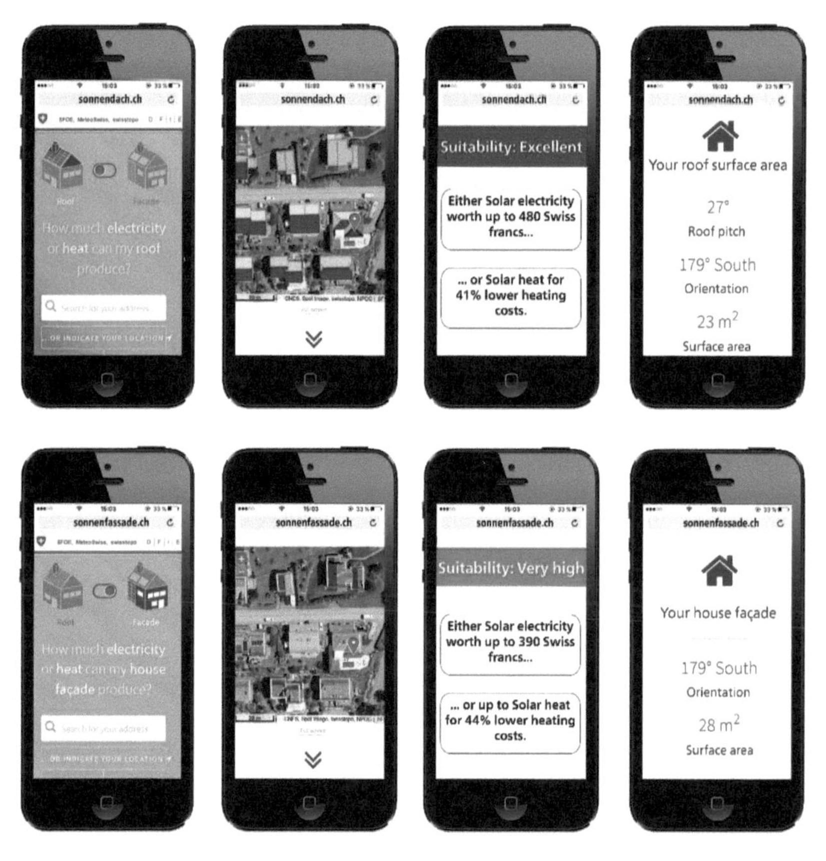


Figure 9.7 Mock-up of the mobile application www.sonnendach.ch (top) und www.sonnenfassade.ch (bottom).

much potential exists for solar energy production in façades, the same calculations are also carried out for the façade surfaces.

#### 9.4.1.2 Easy to use

The application is optimised for mobile devices and is characterised by simple and user-friendly guidance. Each user should be able to learn with minimal effort the potential of his or her own house enabling for the production of electricity and heat. For this reason, the complex 3D data sets were converted back into 2D representations. A user can open the application and reach the desired location (i.e. via the address search, via GPS or by simply navigating in the viewer). Particularly suitable roof and facade surfaces are immediately recognisable by means of different colours in a pictorial representation. In addition, users get information on how much financial monetary value the solar power is worth and what savings they can expect in their heating costs. More technical information on building characteristics, radiation values or solar systems can also be found on the website.

#### 9.4.1.3 Use

This application example shows how lay people can obtain understandable answers to complex questions. Individuals and authorities can use the information from www.sonnendach.ch and www.sonnenfassade.ch as an important basis for planning. In this context, a methodology has already been developed to calculate a cumulative potential for entire municipal areas from the individual potentials. Thanks to the embedding function, municipalities and cantons can easily integrate the data into their own websites and geoportals, thus offering citizens a complete solar register. In contrast to other solar cadastres, the application described above is regularly updated and adapted to the existing conditions. In addition, the federal government ensures that this planning tool is available in all cantons and municipalities and produces comparable data throughout Switzerland.

### 9.4.2 Visualisation

With the Topographic Landscape Model, swisstopo has a powerful documentation of the reality in the form of geospatial data. The contained objects are captured in three dimensions: therefore they are extremely well suited for the accurate modelling and illustration of the real landscape. This allows for the correct presentation of thoroughfares, which cross each other at different levels (bridges, flyovers and underpasses), or trees with a variable expression of size.

In the Topographic Landscape Model, geometries are stored as points, lines and areas in combination with attributes. A street, stored as a line, should appear in the visualisation as an elongated area. Swisstopo has defined geometry and presentation rules for such cases and assigned a colour for each street category. A single tree is stored as a point that represents the top of the tree. For the visualisation, the height of the tree is calculated with the terrain model and then a predefined model of a tree is placed and scaled (see Figure 9.8). These processed data are components for the creation of a virtual reality (VR) or an augmented reality (AR) scene.

Figure 9.8 Street and railway sections on different levels (left) and single trees in different sizes (right) are processed from swissTLM$^{3D}$ with certain geometry and presentation rules.

*Virtual Reality*: The processed 3D-objects from the Topographic Landscape Model play an important role in the virtual reality (VR) scene in addition to the terrain model. However, the richness of objects can very quickly become an obstacle due to the large amount of data. Therefore, there exist several concepts for data reduction (Dörner *et al.*, 2013). Through the use of a visibility analysis from the desired point, objects that are not required can be blinded. A further possibility for data reduction is the implementation of different levels of detail (LOD), where the degree of generalisation increases with the distance to the camera. If the objects cannot be perceived because of the large distance (for example paths), they should be omitted.

The representation or navigation in an extensive scene makes exceptional demands on the hardware and software. A mobile device can reach its limit quickly. As an alternative for locating a specific place in the scene, a pre-rendered 360° panorama is used (see Figure 9.9). The size of the image file is independent from the richness of detail of the scene and requires minimum processing capacity.

*Augmented Reality*: Pokémon Go (https://www.pokemon.com/us/app/pokemon-go/) was a real hype in the topic of augmented reality (AR). Even before that, AR applications for smartphones were available – an example is the wealth of peak finder apps (Figure 9.10), which label the mountain peaks. For the overlay of different layers of information, geometries of the landscape must be determined in the 3D space. The objects of the TLM fulfil this criterion and are suitable for labelling some mountaintops, towns, water bodies or public transport stops. In the mountains, routes are difficult to recognise when distances are far and in winter time, and routes for ski tours are not marked. The terrain view of the camera can overlay the routes and helps users with orientation.

Geometries in AR applications can also be referenced relative to a known object. The predefined object could be a map or an aerial photograph, which is used to superimpose elements of the TLM. Especially as a didactic tool in geography lessons, this combination of analogous and digital content is conceivable.

Figure 9.9 Equi-rectangular panorama from Bellinzona (Castello di Montebello) processed from different swisstopo products and put together into a virtual reality scene.

Figure 9.10 Labelled mountain peaks is the augmented reality function of the iOS mobile application Swiss Map Mobile.

## 9.5 Economic value

The use of 3D geo-information has rapidly developed over recent years. Technological advances have driven this evolution and reduced the costs involved in the procurement and processing of 3D geo-information. An increasing supply of applications is now available and national mapping agencies are actively seeking to transform their data operations and processes into 3D geo-information. On the other hand, an investment in data transformation and processing has to be justified by the additional value, generated by these processes, which raises the question of the value of 3D geo-information. Although this is a very important aspect of the generation, maintenance and use of 3D geo-information, a limited number of papers about the economic value of 3D has been published so far. A few hints on this aspect can be found at Biljecki *et al.* (2015), Ho and Rajabifard (2016), Home (2017), Hubbard (2014) and Altan *et al.* (2013).

Information in itself has limited intrinsic value. The value is derived from its use within a business context or use case. The European Spatial Data Research organisation (EuroSDR) undertook a business case analysis over 12 months in collaboration with 11 European national and regional mapping agencies. In this project, six use cases were selected:

- Forestry management.
- Flood management.
- 3D cadastre and valuation.
- Resilience (civil contingency).
- Asset management.
- Urban planning.

For flood management 3D and 4D geographical information is central for effective planning and risk reduction. There are three processes where the highest value is added, generating the most significant potential socio-economic benefits. First, early warning for emergency services, which simply saves lives. Second, more accurate and reliable risk analysis tools result in better development planning decisions and more appropriate construction. Finally, more accurate and reliable risk analysis tools result in better emergency response planning (including simulations) and therefore in a more effective response.

For 3D cadastre, the analysis covers the uses of geo-information for land registration and administration and land and property valuation. It is doubtless that complex property ownership scenarios are difficult to represent accurately in two dimensions. By using 3D cadastral mapping, lenders can improve the quality of the information that they hold regarding an asset they are lending against. Improvements in information can reduce the amount of liquidity a lender has to hold to secure a loan and can potentially reduce interest rates for the borrower. In addition, 3D cadastral mapping improves the information available to the notary, speeding up the transaction time and associated cost; it also makes the nature of the transaction clear and traceable to other notaries carrying out property searches.

Civil contingency (resilience) planning requires high levels of preparedness and an improved ability of emergency services and government agencies to operate effectively together. The most important benefits of 3D geo-information are more detailed building information (layout, number of floors, etc.), which reduces emergency service response times by improving access decisions (the right resource in the right place at the right time), and better 3D contextual data (e.g. slope, aspect, etc.), which improves localisation of callers and incidents by Emergency Control Room operators.

The advantages of 3D geo-information in asset management, which refers to monitoring and maintaining tangible assets (such as public infrastructure) and intangible assets (such as information), are reducing utility strikes, reducing earthwork volumes on infrastructure projects, and reducing costs of construction.

For Internet of Things (IoT) planning, accurate 3D models of buildings can be used to position sensors in the optimal locations for coverage of the required area and to take into account site access and installation conditions. For telecom planning, where 4G/5G networks require a much greater density of transmitter masts than the previous generation wireless systems, accurate 3D geo-information for buildings, in urban situations particularly, will reduce the costs of planning and implementing such networks.

## References

Altan, O., Backhaus, R., Boccardo, P., Tonolo, F.G., Trinder, J., van Manen, N. and Zlatanova, S. (eds) (2013) *The Value of Geoinformation for Disaster and Risk Management (VALID): Benefit Analysis and Stakeholder Assessment.* Copenhagen: International Council for Science – GeoUnions, Joint Board of Geospatial Information Societies & United Nations Office for Outer Space Affairs.

Biljecki, F., Stoter, J., Ledoux, H., Zlatanova, S. and Çöltekin, A. (2015) Applications of 3D city models: State of the art review. *ISPRS International Journal of Geographical Information*, pp. 1–48.

Chapman, A. D. (2005) *Principles of Data Quality.* Report for the Global Biodiversity Information Facility, Copenhagen.

Dörner, R., Broll, W., Grimm, P. and Jung, B. (2013) *Virtual und Augmented Reality (VR/AR): Grundlagen und Methoden der Virtuellen und Augmentierten Realität* (EXamen.press). Berlin: Springer.

Fonte, C., Bastin, L., Foody, G., Kellenberger, T., Kerle, N., Mooney, P., Olteanu-Raimond, A.-M., and See, L (2015) VGI Quality Control. *ISPRS Annals of the Photogrammetry, Remote Sensing and Spatial Information Sciences, Volume II-3/W5, 2015.* ISPRS Geospatial Week 2015, 28 Sep.–3 Oct. 2015, La Grande Motte, France.

Ho, S. and Rajabifard, A. (2016) Towards 3D-enabled urban land administration: strategic lessons from the BIM initiative in Singapore. *Land Use Policy*, vol. 57, pp. 1–10.

Home, R. (2017) *Cost-Benefit Analysis: Urban Planning Results EuroSDR.* Available at: http://www.eurosdr.net/sites/default/files/images/inline/urban_planning_cost-benefit_analysis_20170316.pdf

Hubbard, D.W. (2014) *How to Measure Anything: Finding the Value of 'Intangibles' in Business. Third edition.* Hoboken, NJ: Wiley.

Jakobsson A. and Giversen J. (2007) *Guidelines for Implementing the ISO 19100 Geographic Information Quality Standards in National Mapping and Cadastral Agencies.* Public report of QKEN eurogeographics.org

Shi, W., Fisher, P. and Goodchild, M.F. (2002) *Spatial Data Quality.* London: CRC Press.

Sulaiman S. and Gudmundsdottir H. (2013) Quality assurance in geodata. Master's thesis, KTH Royal Institute of Technology Stockholm, Sweden.

Zhang, J. and Goodchild, M.F. (2002) *Uncertainty in Geographic Information.* London: Taylor & Francis.

# Index